Crafting Institutions for Self-Governing Irrigation Systems

A publication of
the Center for Self-Governance

Crafting Institutions for Self-Governing Irrigation Systems

Elinor Ostrom

ICS PRESS
Institute for Contemporary Studies
San Francisco, California

This book is a publication of the Center for Self-Governance, dedicated to the study of self-governing institutions. The Center is affiliated with the Institute for Contemporary Studies, a nonpartisan, nonprofit public policy research organization. The analyses, conclusions, and opinions expressed in ICS Press publications are those of the authors and not necessarily those of the Institute, or of its officers, directors, or others associated with, or funding, its work.

Inquiries, book orders, and catalog requests should be addressed to ICS Press, Institute for Contemporary Studies, 243 Kearny Street, San Francisco, CA 94108. (415) 981-5353. Fax (415) 986-4878. For book orders and catalog requests call toll free within the United States: **(800) 326-0263.**

Distributed to the trade by National Book Network, Lanham, Maryland.

Cover designed by MRP Design.

Library of Congress Cataloging-in-Publication Data

Ostrom, Elinor.
Crafting institutions for self-governing irrigation systems / Elinor Ostrom.
p. cm.
Includes bibliographical references.
ISBN 1-55815-168-0
1. Irrigation—Developing countries—Management. I. Title.
HD1741.D44088 1992
631.5′87′068—dc20 91-32946
CIP

CONTENTS

FOREWORD

A self-governing irrigation system is a prime example of a public enterprise in which a segment of society governs itself *for* itself. By agreeing together how water will be apportioned, how responsibilities for maintaining an irrigation system will be allotted, and how such a framework of rules will be enforced and amended to meet changing conditions, water suppliers and users can craft social and political institutions that increase the responsiveness, efficiency, and self-sustaining profitability of irrigation projects.

All too often planners neglect to consult the people most directly involved in an irrigation system's operation when deciding how water should be distributed. Too often, planners fail to ensure that users, who are customarily expected to share in the maintenance of canals, diversion weirs, and other facilities, bear such responsibilities in proportion to their benefits from the system. The result is that both suppliers and customers receive "perverse incentives" to circumvent inflexible regulations and to seek personal advantage in ways that decrease irrigation benefits for fellow users.

Utilizing institutional analysis of irrigation systems large and small around the world, Elinor Ostrom argues that the rules governing how water users interact among themselves and with irrigation managers are just as important to a project's success as are well-constructed engineering facilities.

She describes the workings of several self-organized irrigation enterprises—many of which have functioned for hundreds of years—in which suppliers and consumers have together developed "rules-in-use" that guide the operation of their systems and their individual duties toward them. She explains how such institutions have resulted in an increased willingness of water users to invest labor and resources in the upkeep of irrigation systems—a good indication, she asserts, that they see the benefits of such enterprises as outweighing the costs.

From her analysis Professor Ostrom has compiled a series of "design principles" that can be usefully applied by individuals and communities seeking to craft self-governing institutions—both for irrigation systems and for other common enterprises. These principles provide a fascinating alternative to both "privatization" and bureaucratic management, and offer hope that historically proven community empowerment can guide the crafting of new institutions of self-governance.

Robert B. Hawkins, Jr., President
Institute for Contemporary Studies

PREFACE

This report is addressed to individuals associated with national, regional, and local governmental agencies, donor agencies, indigenous institutions, voluntary associations, farmers associations, and water-user associations, and to analysts interested in irrigation and development. The purpose is to outline an approach to designing irrigation institutions. Supplying and using irrigation water involves a complex set of interrelated activities that are linked over space and time. Attempting to control and use a constantly moving, flowing resource is an endlessly challenging task. If achieved, not only can agricultural productivity be increased, but multipurpose projects can also produce electric power, flood control, navigation, and recreation. The potential for immense destruction is also created whenever large quantities of water are artificially retained.

Most studies of irrigation focus on the creation of physical capital in the form of dams, aqueducts, diversion weirs, and canals. The development of adequate physical capital is, of course, a necessary step in achieving enhanced benefits. But not all technically advanced irrigation systems have produced the projected outcome. Many disappointing investments have resulted from institutional failures. Furthermore, many future efforts will be directed toward improving the performance of existing systems rather than constructing new systems. Thus, while it is essential to understand the physical side of irrigation systems, much of the emphasis in the design of new or rehabilitated systems will be on the institutional side.

This study focuses on social capital in the form of rules and norms of behavior governing how individuals interact. The match of social capital (rules-in-use) with physical capital (engineering works) affects the amount of land that is irrigated, the volume of water provided for productive use, the crop yields achieved, and the distribution of direct and indirect benefits and costs. These can be evaluated using a variety of criteria including (1) sustenance over time, (2) economic efficiency, (3) equity of distribution, (4) accountability of officials, (5) adaptability to changing circumstances, and (6) positive and negative effects on the environment.

The central thesis is that the crafting of institutions is an ongoing process that must directly involve the users and suppliers of an irrigation system throughout the design process. The term "crafting" emphasizes the artisanship involved in devising institutions that both match the unique combinations of variables present in any one system *and* can adapt to changes in these variables over time. Involving users and suppliers directly in this process helps ensure institutions that are well matched to the particular physical, economic, and cultural environment of each system.

This report is a product of the Decentralization: Finance and Management (DFM) Project, sponsored by the Office of Rural and Institutional Development of the Bureau for Science and Technology (S&T/RD) of the U.S. Agency for International Development (USAID). Associates in Rural Development, Inc. (ARD) is the prime contractor for the DFM project under USAID contract number DHR-5546-Z-00-7033-00, with subcontracts to the Metropolitan Studies Program of the Maxwell School of Citizenship and Public Affairs at Syracuse University and the Workshop in Political Theory and Policy Analysis at Indiana University. This report is an annex to an earlier report entitled *Institutional Incentives and Rural Infrastructure Sustainability*, written by Elinor Ostrom, Larry Schroeder, and Susan Wynne. Many of the ideas developed in that report are now presented from the perspective of how they affect the process of crafting irrigation institutions. I am deeply indebted to Larry Schroeder and Susan Wynne for the ideas presented in this report and for the stimulating exchanges we had in preparing the larger study. This report also draws upon my *Governing the Commons* (1990), which treats locally organized irrigation systems as well as other common pool resources in different parts of the world. I am also appreciative of the assistance of Patty Dalecki, Gina Davis, and Sue Jaynes and the comments made on earlier drafts by Roy Gardner, Ronald Oakerson, Vincent Ostrom, Larry Schroeder, Louis Siegel, S. Yan Tang, and James Thomson.

ABOUT THE AUTHOR

ELINOR OSTROM is codirector of the Workshop in Political Theory and Policy Analysis and the Arthur F. Bentley Professor of Political Science at Indiana University, Bloomington. She is the author of *Governing the Commons*, coauthor with Robert Bish and Vincent Ostrom of *Local Government in the United States*, and coeditor with Richard Kimber and Jan-Erik Lane of the *Journal of Theoretical Politics*.

ACRONYMS AND ABBREVIATIONS

DFM	Decentralization: Finance and Management (Project)
ARD	Associates in Rural Development, Inc.
USAID	United States Agency for International Development
GAO	(United States) General Accounting Office
O&M	Operation and Maintenance
NIA	National Irrigation Administration
ARTI	Agrarian Research and Training Institute

CHAPTER ONE

Irrigation, Institutions, and Development

Irrigation development must confront the issues of governance and enlist human and other resources and procedures to arrange appropriate institutions and organizations in addition to appropriate irrigation technologies.

—E. Walter Coward, Jr.
Irrigation and Agricultural Development in Asia

Irrigation Investments and Agricultural Productivity in Developing Countries

The decades between 1950 and 1980 witnessed an almost threefold increase in the total area of irrigated agriculture throughout the world (Cernea, 1985: 23). Dramatic increases in the quantity of foods produced, particularly in developing countries, have resulted from the expansion of irrigated land, the development of new high-yield grain varieties, and the availability of other agricultural inputs. In many countries, such as India, Indonesia, Pakistan, the Philippines, Sri Lanka, and Thailand, the most important factor affecting the quantity of rice produced has been the amount of land subject to irrigation (Dhawan, 1988: 13–15; Carruthers, 1988: 9; Madduma Bandara, 1977: 298–301).[1] The spread of irrigation has "contributed between 50 and 60 percent of the massive increase in agricultural output of the developing countries from 1960 to 1980" (Crosson and Rosenberg, 1989: 130).

Expanded agricultural production in developing countries outside Africa has resulted from massive investments in large-scale irrigation projects by donor agencies and host countries, in addition to investments in new agricultural inputs and techniques.[2] The World Bank alone provided over $11 billion in loans for irrigation and drainage projects between 1947 and 1985 and another $7.5 billion for area development projects that frequently included substantial irrigation activities.[3] Thirteen percent of the loans issued by the Asian Development Bank during the 1970s were related to irrigation projects (General Accounting Office, 1983: 2). Some individual projects were very costly. The Rahad scheme in the Sudan, for example, cost donors and the government of the Sudan $400 million.[4] The enormous Mahaweli project in Sri Lanka was planned to develop or improve water supply for 900,000 acres of land and for over 200,000 new settlers (Jayawardene, 1986: 79). Bilateral aid agreements provided grants and import support to the Mahaweli project of at least $365 million (in 1982 U.S. currency), for which no repayment was due (Ascher and Healy, 1990: 100).

The Lack of Sustainability of Many Large-Scale Irrigation Projects

Even though the massive investments in irrigation have generated higher agricultural yields,[5] many large-scale irrigation projects have not been sustainable; that is, after the project was completed, the net flow of costs exceeded the net benefits. Failures occur when costs exceed benefits. One way the World Bank and other donors determine economic sustainability is by assessing whether the economic rate of return is at least equal to, if not greater than, the opportunity cost of capital (Cernea, 1987: 3). By this standard, many large-scale irrigation projects have generated disappointing operational results (see, for example, International Bank for Reconstruction and Development, 1985). The benefit-cost evaluation of the original Gal Oya scheme in Sri Lanka, for example, showed that discounted costs exceeded discounted benefits by 277 million rupees ($51.25 million in 1957 U.S. currency) (Harriss, 1984: 318). In many other projects, actual costs have so exceeded projected costs that economic sustainability is unlikely. The costs of the completed irrigation works for the Jamuna Irrigation Project in India, for example, amounted to 69.80 million rupees ($9.07 million in 1969 U.S. currency), as contrasted with the estimated project cost of 39.60 million rupees ($5.15 million in 1969 U.S. currency) (Ascher and Healy, 1990: 147).

The lack of an infrastructure for sustainable irrigation in many developing countries has been attributed to many causes. One problem has been the tendency for initial benefit-cost analyses to be unrealistically optimistic (Pant, 1984: xvii). Underlying that optimism are several systematic biases that tend to occur in initial planning for major irrigation projects. The area to be irrigated (or to receive water in a second planting season) is frequently much larger in the projected plans than is realized in practice. For instance, the area *actually* irrigated in the Uda Walawi scheme in Sri Lanka covered only one-third of the area projected when the project was funded. Much of the land that planners presumed would produce two crops has produced only a single crop after project water was made available. In the Jamuna project, only 31 percent of the targeted service area was brought under irrigation by 1974 when the main headworks, diversion works, and distribution canals were completed (Ascher and Healy, 1990: 143).

Another systematic problem leading to overly optimistic benefit-cost ratios is overestimation of the agricultural yields to be obtained. Agricultural yields obtained *after* project construction have sometimes been lower or more variable than anticipated. Mehra (1981) reports that the variability of crop yields after the construction and operation of major irrigation systems in India increased rather than decreased. Levine (1980: 55) reports that Iranian irrigators using a traditional system with minimal facilities had been able to achieve water-use efficiencies (water delivered to field inlets as a percentage of water supplied to distribution intakes) of approximately 25 percent before the construction of the Dez Pilot Irrigation Project. This project was "a comprehensive system, with a full range of controls, measuring structures, organizational structure, and all the other accoutrements of a large modern system." Six years after the Dez project was completed, the average water-use efficiency in the area had fallen to between 11 and 15 percent. Bromley (1982) reports similar reductions in water-use efficiencies for major projects throughout Asia.

Another major reason that irrigation projects have lacked sustainability is underinvestment in recurrent costs associated with the operation and maintenance (O&M) of the systems. A World Bank study of forty-eight recently constructed irrigation projects showed that O&M expenditures were at the level agreed upon with the host government in only half the projects. "Clearly many were already well on their way to becoming fashionable rehabilitation projects" (Carruthers, 1988: 9). In 1983, the U.S. General Accounting Office (GAO) conducted a survey of USAID-funded irrigation projects in Indonesia, Sri Lanka, and Thailand and found many of them in poor condition because O&M activities had not been undertaken (GAO, 1983). The same report found that each of these countries delayed routine maintenance until deterioration of the systems was extreme enough to require rehabilitation, largely funded by donor agencies. The GAO drew the following conclusions:

> A primary reason for this is inadequate funding of the day-to-day regular operation and maintenance, or recurrent costs. . . . O&M funds must come from the host governments, the system users, or donors through additional or redirected assistance. Host-government budgets have been inadequate and user fees have not been collected regularly. Donors normally restrict their financial involvement to design and construction and view operation and maintenance as a recipient country responsibility. (GAO, 1983: 6)

The report contained the following specific findings:

- At Indonesia's Luwu Irrigation Project, it was evident that no routine maintenance was being performed.

- At Indonesia's Rural Works' subproject sites, we found heavy erosion damage to canal banks. In addition there was siltation and weed growth which eventually can restrict water flows. There were signs of vandalism at all of the Sederhana subproject sites visited.

- At Sri Lanka's Mahaweli Irrigation Project, we saw many examples of poor operation and maintenance, including weed growth in canals and more evidence of farmer vandalism.

- In Thailand, at all three irrigation projects we saw silt and weeds in the canals and holes and cracks in the concrete canal linings. Small, unattended problems gradually grow until major repairs are needed. (GAO, 1983: 6–7)

Perverse Incentives

Underlying all these problems are a variety of perverse incentives. These lead to the overestimation of benefits to the producers and consumers of agricultural products, the underestimation of the costs of sustaining irrigation projects, and the actual underinvestment in operation and maintenance activities on irrigation projects in developing countries. Project engineers, for example, face strong pressures to focus on the design of physical works while ignoring social infrastructure and to focus on larger, rather than smaller, projects. Farmers on large-scale projects face perverse incentives associated with their lack of control over water availability and substantial temptations to refrain from contributing resources to maintenance.

The initial plans for many of the major irrigation projects in developing countries have focused almost exclusively on engineering designs for the physical systems. Distribution of water to farmers and subsequent maintenance were frequently not addressed (Chambers, 1980; Bottrall, 1981).[6] In the Sri Lankan Mahaweli project, planning focused exclusively on the physical systems and ignored organizational questions.

> It was assumed by the planners that the farmers in each turnout would, on their own, organize themselves for the equitable distribution of the water allocated to them. They also assumed that the farmers would maintain their field channels and irrigation structures on their own. (Jayawardene, 1986: 79)

The engineering bias rapidly triggers perverse incentives for irrigators. An evaluation of the Mahaweli project five years after completion found that only half the farmers being served received water through authorized outlets of canals (Corey, 1986). The other half obtained water through illegal diversions or from drainage of other fields. Instead of following regular rotation systems, farmers blocked and unblocked the ditches and outlets, trying to get more than their authorized shares. At times, upstream irrigators were able to obtain the full flow of an irrigation canal. Corey described one incident in the following way:

> In one case, an unauthorized breach was observed to be taking the entire supply of water from a ditch. The downstream farmer said he was not able to obtain water to irrigate his paddies even though he had appealed to the farm leader. When asked why he did not close the breach himself, he said he was afraid of being assaulted by the man who had made the breach. When the farm leader was asked why he permitted this situation to exist . . . he said he was afraid to take further action on his own initiative for fear of being "hammered" by the offending farmer. (Corey, 1986)

Such incidents occur frequently on large-scale irrigation projects. "Common practices include constructing illegal outlets, breaking padlocks, drawing off water at night, and bribing, threatening, or otherwise in some way inducing officials to issue more water" (Chambers, 1980: 43). The initial lack of attention to such problems leads to uncertainties in water deliveries and water rights. With such uncertainties, farmers are less willing to try new seed varieties or adopt the associated cropping schedules. Unpredictable availability of water also induces farmers to avoid investments in construction and maintenance of field channels.

One major bias that has characterized much of the planning for irrigation projects in developing countries is an assumption that large projects produce the most benefits. Considerable evidence, however, indicates that smaller projects—minor irrigation works—have a higher potential for substantial returns than larger projects. A decade ago, Roy (1979) assessed the progress of the Green Revolution in northern India and identified small irrigation systems as the key factor leading to the most impressive increases in productivity. After a sweeping analysis of irrigation experiences in Africa, Moris and Thom (1990) conclude that higher returns are possible in small-scale projects.

Many factors contribute to the support of large irrigation projects. Farmers themselves may favor large-scale projects because they believe that these projects will be provided to them at low costs. Water from large-scale projects is frequently highly subsidized (if not entirely "free"). Farmers' support for low-cost water is quite understandable. Projects that support credit to farmers for the renovation of small-scale projects place the risk on the farmer rather than on the donor agency or host government. Although the hope of obtaining free benefits frequently leads farmers to support large-scale projects, farmers will support small projects if other types are not foreseen.

The settlers on some large irrigation systems have so little choice about which crops to plant, how to use the land, which inputs to purchase, and when to sell crops, that yields are consistently lower than predicted. Settlers commonly attempt to find work outside the project rather than devote their efforts to increasing agricultural yields. For example, the massive (882,000 ha) Gezira scheme in the Sudan delimited 102,000 tenancies in which tenants were given almost no independent decision-making authority over the land's use (Barnett, 1977). Until 1980, a joint-account system was in use on this and most other irrigation schemes in the Sudan. With the joint-account method, a disproportionate share of system operating costs (which included costs for growing crops other than cotton) was deducted from cotton revenues. Tenants were then allocated a return based on a set formula regardless of their own productivity. With these perverse incentives, it is little wonder that the level of cotton productivity steadily declined; tenants were inclined to grow crops other than cotton and to gain employment outside the scheme altogether. Presently, even after the adoption of an individual account that pays tenants for the amount of cotton harvested from their assigned tenancy, more than half the labor requirements on the project are met by migrant labor (Plusquellec, 1990: 33).

In developing countries, politicians may derive more electoral support by announcing a new project that will cover a large area serving many individuals than by announcing a credit program that will help many small-scale irrigation systems to improve their facilities or expand their service areas by small amounts. Agency officials are professionally encouraged to promote projects that deliver water to as many farmers and as much land as possible. This encouragement results in agency support for large projects and a tendency to exaggerate the actual area served by many large-scale projects in official records.

The Need to Organize the Farmers

The persistent problems with the design, construction, operation, management, and use of irrigation projects have led donors and national governments to reevaluate the emphasis on engineering in irrigation planning and to stress the importance of organizing farmers to make the most effective use of the capital investment. The Asian Development Bank was among the early advocates of farmer organization:

> The success of an irrigation project depends largely on the active participation and cooperation of individual farmers. Therefore, a group such as a farmers' association should be organized, preferably at the farmers' initiative or if necessary, with initial government assistance, to help in attaining the objectives of the irrigation project. Irrigation technicians alone cannot satisfactorily operate and maintain the system. (Asian Development Bank, 1973: 50)

A decade later, USAID sponsored an evaluation team to undertake a worldwide, comprehensive assessment of irrigation projects. The team concluded that "too often the effort begins with construction to the original blueprint, with complete neglect of the social, institutional, and managerial dimensions" (USAID, 1983: 90). The team called for organizing farmer participation in allocating, financing, and maintaining major irrigation systems.

At the same time, the 1983 GAO study pointed to the need for establishing farmer cooperation on most major irrigation projects, given the great numbers of very small farmers served by irrigation projects in developing countries. "Without close cooperation," the GAO report argued, "some farms will receive more water than needed, others will do without, and routine maintenance will not be shared among all those receiving irrigation benefits" (GAO, 1983: 36). This report also urged the establishment of water-user associations that could undertake most of the routine maintenance on distributary canals, as well as articulate the needs and interests of the farmers to project officials. In the 1990s, donor agencies are concerned that future irrigation projects involve major efforts to organize farmers for the development of effective rotation or other allocation plans and for the maintenance of the field-level irrigation works.

Organizing farmers is now stressed in documents written by donor agencies, host governments, and development scholars (see Brown and Korten, 1989). Some notable success stories have been written. The establishment of effective farmer organizations on the San Lorenzo Irrigation Project in Peru helped increase agricultural productivity substantially. The farmers there have undertaken responsibility for water allocation and canal maintenance. The upkeep of the system has thereby been enhanced. Project benefits continue to be sustained long after the project was completed (Cernea, 1987).

Similar successes were achieved by the Mexico Third Irrigation Project (Cernea, 1987). This project involved a successful revitalization of previously existing, but relatively inactive, *ejido* organizations. Membership in the *ejidos* continued to grow steadily after project completion. More than five years after the official project was completed, farmers who were members of the *ejidos* had earned a threefold increase in average farm income, were undertaking new entrepreneurial functions, and were sustaining their previous activities. Unfortunately, not all government-owned systems in Mexico have been as successful as the Mexico Third.

In addition to government-owned irrigation projects in Mexico, there are around 13,700 farmer-owned irrigation systems, called *Unidades de Reigo,* that were responsible for irrigating more than 1.5 million hectares in 1982. The *Unidades* are "structured and operated as Irrigation Communities (they own the infrastructure, operate it as a common property resource, charter the CEO, and duties and benefits are tightly integrated)" (Hunt, 1990: 149). Given these institutional differences between government-owned and farmer-owned systems, participation in farmer-owned systems is rarely problematic. "There is no question about the presence of farmer participation in these systems: The farmers manage the system, perform maintenance, and pay for all the O&M" (Hunt, 1990: 150).

Plusquellec (1989) describes the successful efforts of the Colombian government to transfer management responsibilities to water-user associations on a gradual basis. A medium-sized project in the Coello district—one of the first projects to be turned over—has been successfully managed by a water-user association since 1976. The system is well maintained. The costs of operation and maintenance are modest ($35 per hectare in 1989 U.S. currency) and fully covered by a water charge collected from *all* farmers served by the district (Plusquellec, 1989: 4). The experimental program successfully adopted within the National Irrigation Administration of the Philippines has also demonstrated that the active participation of farmers in the early stages of project planning and the mobilization of those resources needed to reconstruct physical works can enhance long-term sustainability (Korten and Siy, 1988; see discussion in Chapter 5).

In an evaluation of major development projects demonstrating long-term sustainability, the World Bank stressed the role of successful farmer organizations:

> A major contribution to sustainability came from the development of grass roots organizations, whereby project beneficiaries gradually assumed increasing responsibility for project activities during implementation and particularly following completion. . . . Where grass roots organizations thrived there were certain distinct qualities inherent in their growth and in their relationships to project activities. These included some form of decision-making input into project activities, a high degree of autonomy and self-reliance, a measure of beneficiary control over the management of the organization, and the continuing alignment of the project activities with the needs of the beneficiaries. (IBRD, 1985: 35–36)

In some regions, farmers have been organized for long periods of time, and existing farmer organizations are quite effective. For example, the most effective water-user associations visited by a GAO team in 1983 were the Balinese *Subaks* in Indonesia.

> Their irrigation systems appeared to be well maintained and in excellent condition. The *Subaks* had, in most instances, designed and constructed their own systems; the religious and ethnic structures were an important part of the association; each *Subak* had a strong organizational structure; and fees were collected to help operate and maintain the system. (GAO, 1983: 38)

The Balinese *Subaks* have been organized over the centuries by the farmers themselves without guidance from central authorities. Although general principles of organization are used by all *Subaks*, the specific rules used in each *Subak* vary to cope with the specific problems faced in governing each individual system (Geertz, 1980).

Strong indigenous irrigation institutions also exist in the Philippines and in Nepal and have outstanding records of sustainability (see Uphoff, 1986; Coward, 1980; Pradhan, 1989a; Sampath and Young, 1990).

Although organizing farmers is now acknowledged to be a key step in successful irrigation projects, many projects are not as successful in stimulating grass-roots organizations as those described previously. On the Sriramasagar Project in India, for example, government officials met in the mid-1970s with farmers on thousands of outlets to create Pipe Committees that could take responsibility for water distribution, rule enforcement, and conflict resolution. Although farmers came to the initial meetings in considerable numbers, no real organization took root (Singh, 1983). On the Mula Project in Maharashtra, *Pani Panchayats* were reportedly established on 24,000 hectares by 1985 (Patil, 1986, cited in Chambers, 1988: 90). But these paper organizations were not much more than "mere euphemisms" for the meetings held by project authorities to inform farmers of administrative decisions. In reviewing the reasons for failed efforts to organize the farmers, Chambers concludes that farmers cannot be organized through "persuasion or fiat" and "will only participate if they see some gain from doing so" (Chambers, 1988: 90; see also Gillespie, 1975).

The effort to develop farmer organizations has frequently consisted of central officials designing the skeletal structure of the type of organization they will formally recognize. This design is then viewed as a predetermined "blueprint" for how farmers will organize themselves. On some projects, officials have ignored preexisting irrigation associations and have recognized only their own newly established farmer organizations (see discussion in Coward, 1985: 33–36). On other projects where efforts have been made to organize farmers, farmers meet and elect the officials they are requested to elect, but any further organization is thwarted.[7] Farmers resist efforts to develop water allocation procedures and refuse to participate in the maintenance of the field canals. Consequently, officials perceive farmers as intransigent, irresponsible, and irrational. The failure of these projects to meet predicted benefit levels is blamed on the farmers rather than on engineering design or on the lack of effective institutional development.[8]

Irrigation in the Twenty-first Century

Although irrigation investments in the latter half of the twentieth century have frequently lacked sustainability, they have helped to produce the spurt in agricultural yields needed to avert a massive shortfall of food to feed the growing population of the developing world. Population levels have steadily increased since 1950, but agricultural productivity has increased even faster. Unless far more effective irrigation institutions are designed in the future, it is unlikely that increased agricultural production will continue to outstrip increased population levels in developing countries. This is the case for several reasons:

- The least expensive irrigation sites have already been developed in most of these countries. The costs of new investments in large-scale projects tend to rise faster than farm produce prices.[9] Thus, the rate of new irrigation water made available to farmers from new large-scale projects will slow considerably (Yudelman, 1989: 66, 74; Dhawan, 1988: 240; Moris and Thom, 1990: 39–40).

- Maintaining current irrigation projects at full operating capacity will become more expensive given the lack of maintenance provided during the past several decades (Yudelman, 1989: 68).

- Further dramatic increases in the yield potentials of crops are somewhat unlikely.

- Many environmental problems resulting from past investments in irrigation are now becoming apparent, and opposition to the construction of new large-scale irrigation projects is growing (Yudelman, 1989: 69–73; Moris and Thom, 1990: 33–39; Kaye, 1989: 16).

As a consequence of these problems, there will be fewer investments in new irrigation projects made in the future than have been made in the last several decades.[10] To get more irrigation water to the farmer at the times and places that are most important for increasing agricultural yields, major improvements in the operation and maintenance of existing irrigation systems must be made. A study of forty irrigation service areas in Pakistan, for example, found that "5 million acre-feet of scarce water could be saved in the Punjab and Sind for field application simply by proper maintenance of the local community watercourses" (Freeman and Lowdermilk, 1985: 107). Although some improvements in the operation of existing irrigation systems can come from better physical structures, particularly control structures, the key problems relate to the incentives facing officials and farmers. As long as few individuals are motivated to operate and maintain irrigation systems effectively, actual agricultural yields produced in areas served by large-scale irrigation projects will continue to be disappointing.

The Importance of Institutional Design and Social Capital

Over the next several decades, the most important consideration in irrigation development will be that of *institutional design*—the process of developing a set of rules that participants in a process understand, agree upon, and are willing to follow. An embedded institutional design is a form of *social capital*, defined by James Coleman (1988) as the aspects of the structure of relationships between individuals that enable them to create new values. *Physical capital* is embodied in the tools, machines, and physical works that enable individuals to produce goods and services. *Human capital* is created by "changes in persons that bring about skills and capabilities that make them able to act in new ways." *Social capital*, on the other hand, is created "through changes in the relations between persons that facilitate action."

> If physical capital is wholly tangible, being embodied in observable material form, and human capital is less tangible, being embodied in the skills and knowledge acquired by an individual, social capital is less tangible yet, for it exists in the *relations* among persons. Just as physical capital and human capital facilitate productive activity, social capital does as well. For example, a group within which there is extensive trustworthiness and extensive trust is able to accomplish much more than a comparable group without that trustworthiness and trust. (Coleman, 1988: s100–101)

Designing institutions involves creating new forms of relationships between individuals. The process of institutional design is quite different from that of engineering design. As experience with organizing farmers over the last several decades has shown, simply giving individuals organizational blueprints is not equivalent to changing the incentives and behavior of those individuals. Nor is the problem simply that of organizing farmers. Many perverse incentives face design engineers, construction firms, and the officials responsible for operating and maintaining irrigation systems. Both the failure to achieve project sustainability and the failure to organize farmers illustrate a pervasive lack of understanding as to how effective institutions are crafted over time.

This report outlines an approach to the design of irrigation institutions that is useful to officials in donor agencies, host governments, and other agencies or organizations involved in the design, operation, and maintenance of irrigation projects in developing countries. The crafting of irrigation institutions is an ongoing process that must directly involve the users and suppliers of irrigation water throughout the design process. Instead of designing a single blueprint for water-user organizations to be adopted on all irrigation systems within a jurisdiction, officials need to enhance the capability of suppliers and users to design their own institutions. Involving suppliers and users directly will help ensure that development institutions are well matched to the particular physical, economic, and cultural environment of each system.

Although this approach presumes that the participants need to be involved in the design process, it does not presume that good institutional designs spring up naturally as the result of spontaneous organization. Government officials and donor agencies can and should play an active role in enhancing the design process and monitoring the results. The role proposed for central governmental officials and for donor agencies is, however, quite different from that proposed by earlier studies that called for the creation of many user organizations based on the same institutional design.

Proposals for reform are presented at the end of Chapter 5. But first, Chapters 2 through 4 describe the general approach used here to the institutional analysis of irrigation systems, since it differs significantly from many of the current approaches to the study of development processes. Chapter 2 focuses on the significance of viewing institutions as "rules-in-use" rather than as paper organizations created by formal legislation without participation by those affected. Chapter 3 discusses the process of crafting institutions. Chapter 4 presents the design principles derived from an intensive study of several long-enduring self-organized irrigation systems. Finally, Chapter 5 focuses on the problems of applying these design criteria in efforts to improve both government-owned and farmer-owned irrigation systems.

Notes

1. The introduction of high-yield varieties has not always been associated with higher yields (see Byrne, 1986). For a discussion of agricultural technology see Groenfeldt and Moock (1989).

2. The relationship of labor, land, and other agricultural inputs in most of Africa is considerably different from that in most other developing regions. Land is abundant and labor is relatively scarce throughout most of Africa. Efforts to expand agricultural production through massive irrigation projects in Africa have been far less successful than in Asia (Moris and Thom, 1990; Binswanger and Pingali, 1988).

3. Computed from the annexes to Yudelman (1985).

4. The Rahad project is one of the most centralized large-scale projects undertaken with donor funding. A project evaluation noted the following:

> From recruiting and settling tenants to their possible eviction due to failure to meet contract conditions, the corporation maintains strict authority. It provides all agricultural inputs and markets and processes the cotton production. More than this, through controlled monitoring and sanctions it supervises what decision-making is to occur on each tenancy and assesses all costs against profits. (Benedict et al., 1982: 5)

The evaluation concluded that the low production efficiency of the project resulted from the "top-down management structure" that sacrificed critical knowledge from practicing farmers (Benedict et al., 1982: 17).

5. "World grain production increased from 620 million tons in 1950 to 1,660 million tons in 1985, and the average yield per harvested hectare climbed from 1.1 tons to 2.6 tons" (Wolf, 1986: 9).

6. Freeman and Lowdermilk (1985: 96) provide the following overview of the design process:

> In most large-scale systems, especially in Asia, the upstream control systems are designed without regard to the problems faced by farmers in securing local control over irrigation water. Engineers traditionally have provided a transport system for water via rivers, canals, reservoirs, and diversion structures. They have assumed that if water flowed in the general direction of command areas, good water management at the local level would evolve automatically simply because it was needed.

7. David Groenfeldt describes two such systems in which there are "farmer leaders" but no "farmer organizations."

> In Kalankuttiya, there is a farmer representative who is elected every three years; however, many farmers don't know who he is, and those who do know rarely communicate with him. In Dewahuwa, a farmer representative is selected by farmers to coordinate the farmers within a turnout group. However, a turnout group can have as many as 50 farmers who may or may not be located in the turnout, may or may not be owners of the land they cultivate, and may or may not know each other on a personal level. Farmer representatives for each turnout meet periodically with irrigation officials, but it would be inaccurate to say that they represent a group consensus among turnout farmers. (Quoted in Colmey, 1988: 4)

8. The frequency with which farmers are blamed for the failure of irrigation projects inspired the following satirical characterization of the six phases of irrigation project development:

> The first phase is the designers' high enthusiasm and publicized expectations. Second comes *disillusionment,* when the implementors discover that the designs are sorrowfully inadequate. The third phase is one of panic, when the operational staff discovers that the system will not operate as designed. Fourth comes the search for the guilty, characterized by a round robin of blame among designers, implementors, operators, and extension workers. Naturally, the fifth phase consists of blaming the innocent—that is, the farmer who had nothing to do with designing, implementing, operating, or extending the system. Thus, reports sadly conclude that ignorant and stubborn farmers remain set on destroying structures, stealing water, and creating all kinds of other problems and in general will not

> cooperate with well-meaning project authorities. Phase six is the time for praise; if a system works at 40 to 50 percent of design efficiency the praise and honor for the success go not to the planners, engineers, technicians, or the farmers, but the politicians. (Freeman and Lowdermilk, 1985: 91–92)

9. Yudelman (1989) reports that "discussions with [World] Bank Staff indicate that average costs per additional hectare irrigated by some new projects have increased from less than $1,000 to over $5,000, and in a few cases, have even reached $10,000."

10. Ian Carruthers (1988) summarizes a recent FAO report that estimated the rate of growth of irrigated agriculture was 5 percent per annum in the period of 1965 to 1975 and that it fell to 1.5 percent per annum during the next decade.

CHAPTER TWO

Institutions as Rules-in-Use

The concept of institutions is crucial in analyzing why many institutions established for the supply and use of irrigation water create perverse incentives leading to the nonsustainability of irrigation projects. In the development literature the term "institution" can refer to a specific organization in a particular country, such as the Department of Irrigation; it can describe established human relationships in a society, such as family structure (the institution of the family); or it can denote the rules that individuals use to order specific relationships with one another. This paper uses the term "institution" in the third sense: an institution is simply the set of rules actually used (the *working rules* or *rules-in-use*) by a set of individuals to organize repetitive activities that produce outcomes affecting those individuals and potentially affecting others. Hence, an irrigation institution is the set of working rules for supplying and using irrigation water in a particular location.

Working rules are used to determine who is eligible to make decisions in some arena, what actions are allowed or constrained, what procedures must be followed, what information must or must not be provided, and what costs and payoffs will be assigned to individuals as a result of their actions (E. Ostrom, 1986). All rules contain prescriptions that forbid, permit, or require some action or outcome. Working rules are those actually used, monitored, and enforced when individuals make choices about the actions they will take in operational settings or when they make collective choices (Commons, 1957). Enforcement may be undertaken by those directly involved, by the agents they hire, by external enforcers, or by a combination of these. Rules are useless unless the people they affect know of their existence, expect others to monitor behavior with respect to these rules, and anticipate sanctions for nonconformance. In other words, working rules must be common knowledge and must be monitored and enforced.

Common knowledge implies that every participant knows the rules, knows that others know the rules, and knows that others also know that the participant knows the rules.[1] Institutional rules must be known, understood, and followed (in a high proportion of relevant instances) by more than a single individual. By contrast, prescriptions that an individual imposes on personal actions without expecting others to impose the same prescriptions on their own actions are norms or moral strictures and are not included in this definition of rules.

Working rules may or may not closely resemble formal laws that are expressed in national legislation, administrative regulations, and court decisions. A system that is governed by a "rule of law" is one in which formal laws and working rules are closely aligned and enforced. Although formal laws are often a major source of the working rules used in many irrigation systems, particularly when conformance to these laws is actively monitored and sanctioned, this is not always the case. In some irrigation systems, the working rules used by irrigators differ considerably from legislative, administrative, or court regulations (see, for example, Wade, 1988). The difference between working rules and formal laws may involve no more than filling in the lacunae left in a general system of law. More radically, working rules may assign *de facto* rights and duties that are contrary to the *de jure* rights and duties of a formal legal system. Communities of irrigators may use their own institutional arrangements to reach accommodations at variance with the formal rules established by edict. Because rules-in-use are not equated with written laws or regulations, rules-in-use are not directly observable phenomena. It is the *activities* organized by rules that can be directly observed.

Visible Activities and Organizations, and Invisible Institutions

An engineer designing a new irrigation system is observed working at a drafting table preparing drawings or blueprints. A water distributor is observed on a canal opening or closing valves and farm-gates to allow the water to flow in predictable ways. A farmer is observed clearing weeds from a field channel. Are these activities organized by a set of rules? If these activities are related to irrigation works that jointly affect a group of individuals (rather than to a project confined to the land of a single individual), then the answer is almost certainly yes. The kind of training the engineer has received before undertaking this activity, the way in which the engineer was given the assignment to design the system, the type of works considered, the objectives and constraints on the design process, and the way that the engineer will be rewarded for the design are all affected by the rules used in a particular setting. Similarly, rules-in-use will affect how the water distributor obtains his or her position, how the water is distributed, how the distributor obtains money (or other resources) from an employer or from the farmers, which channels are cleared by farmers, and at what times they are cleared.

Most of the rules affecting the design engineer (such as those related to the engineer's prior training) may conform to the formal administrative procedures of a particular ministry. If these formal requirements are consistently waived for individuals closely related to important governmental officials, however, the rules-in-use differ from the formal requirements. Other rules affecting the engineer's work may not be specified in formal law; instead, they will have evolved *in situ*. For example, if external donor assistance will be requested to help finance the construction of new irrigation systems, the ability to maximize the number of individuals who could potentially be served by these systems may be an explicit or implicit design criterion used in evaluating the engineer's work. Thus, the design criterion affects the engineer's incentives.

Similarly, the water distributor's activities are likely to be affected by a diverse set of formal laws or administrative procedures as well as many shared understandings that have evolved locally about payoffs for activities. Some of these understandings may stand in direct opposition to formal legislation or administrative procedures. Accepting bribes from local farmers for delivering water to them is usually forbidden in the formal procedures of irrigation agencies. In some agencies, however, payment for water delivery is so routine that the exact price for various types of services performed is well known to all farmers and to most officials working in the agency (see Wade, 1982a, 1982b). Finally, the observed canal-cleaning activities of the farmer may be the result of an agreement with one or two neighbors, in which each will clean the canal adjacent to his or her own land; this may be part of a complex set of agreements embedded in the working rules of a farmer association.

The activities undertaken by the engineer, the water distributor, or the farmer may be organized with respect to the rules of a particular organization such as an irrigation department or a water-user association. Organizations, like activities, are frequently easier to observe and measure than the rules-in-use of an organization. Many activities, particularly those related to irrigation, are the result of multiorganizational arrangements. The water distributor may be trained by an irrigation department but paid by a water-user association, as in some systems in Taiwan, for example (Levine, 1980). Most large-scale irrigation systems involve the activities of several different organizations, including international donors, national governments, private contractors, and water-users' organizations.

Rules-in-use are similar to knowledge-in-use in the sense that they are invisible to direct observation. For example, we can observe an individual's record of formal education to learn about his or her course of study and the number of years of education completed; however, we cannot directly observe the actual knowledge that an individual uses in undertaking activities, nor can we know the exact source of this knowledge.

Determining what rules are in use in a system is also similar to determining knowledge-in-use. To evaluate the level and type of knowledge an individual uses, we need to interview that individual and also observe how the individual performs various tasks. Similarly, to ascertain what rules a *set* of individuals uses, we need to interview those individuals and observe how they perform activities. Asking questions and administering tests (such as achievement tests) to determine the level and type of knowledge possessed by individuals are essential but imperfect measures of knowledge-in-use. Better evaluations are made by watching how individuals solve particular problems. Similarly, the task of determining the rules used by the suppliers and users of an irrigation system cannot be completely determined by an outsider asking questions. More valid judgments come from long-term observation of how the individuals supplying and using irrigated water undertake organized activities.

In some systems, it may be possible to observe events or markers that directly result from behavior that conforms to rules-in-use. Property rights to water, for example, are often physically manifested in the weirs used on irrigation systems to allocate water to channels serving particular farmers (Coward, 1980). In Nepal, for example, the property rights of different participants in some hill irrigation systems are implemented through the use of wooden proportioning weirs called *saachos* to allocate water automatically (see Pradhan, 1989a). The weirs operate to distribute water in conformance with specified property water rights. Here physical markers indicate a set of agreements about who should receive what proportion of the flow of an irrigation system.

On the other hand, the presence of physical markers associated with particular rules may give false impressions. In the early 1970s, considerable pressure was exerted by government officials in many regions of India to establish rotational water systems similar to the traditional *warabandi* systems used since the nineteenth century in northwest India and Pakistan (Chambers, 1988: 92). *Warabandi* boards were posted to provide general information about the day of the week and time when water was supposed to be allocated to a particular farmer. Casual inspection would seem to indicate that an allocation rule involving strict rotations was in force. In some of these systems, however, the boards only signified a failed effort by outsiders to impose foreign rules on local farmers. Two out of five farmers served by systems supposedly using the "new *warabandi*" rules could not tell a survey taker the day and time of their *own* turn. One-fourth of the respondents could not even explain how a *warabandi* distributional system worked (Chambers, 1988: 93).

The difficulty of observing institutions frequently results in two errors. The first is the assumption that the rules-in-use are always the same as formal laws or procedures. The second is the assumption that no institutions exist except for those that have been formally created through governmental actions. Both errors reflect a lack of understanding of how to create, maintain, and use social capital.

The first error—assuming that institutions are equivalent in practice to what has been written in formal legislation—leads to misplaced confidence in the effectiveness of changing behavior by changing formal law. In a polity characterized by a high conformance to legal prescription, working rules will fill in the details of general legislation. In a system where a rule of law does not prevail, working rules may vary substantially from legislation—particularly legislation drafted by officials located in distant capital cities. If analysts erroneously assume that individuals automatically learn about, understand, and use all the rules contained in formal laws, the development strategy adopted will focus primarily on the activities of central legislatures and administrative agencies, with little attention to what actually occurs in the field.

The second error—assuming that no institutions exist unless created by governmental action—may lead to actions that destroy existing institutions. Coward (1985) reported that farmers in the Philippines, who had already invested many years in crafting local institutions, discovered that new irrigation projects presumably designed "for their benefit" were destructive of the very institutional capital they had worked so hard to create.

Why Do Institutions Matter?

If institutions are invisible, why do they matter? There are several reasons. Institutions shape the patterns of human interactions and the results that individuals achieve. Institutions may increase the benefits from a fixed set of inputs; conversely, they may lower efficiency so that individuals have to work harder to achieve the same benefits. Institutions shape human behavior through their impact on incentives.

The concept of *incentives* involves more than just financial rewards and penalties. Incentives are the positive and negative changes in outcomes that individuals perceive as likely to result from particular actions taken within a set of working rules, combined with the relevant individual, physical, and social variables that also impinge on outcomes. Chester I. Barnard, an administrative practitioner of great skill and a cogent observer of organizational life, provided a relatively comprehensive overview of the concept of incentives. He summarized incentives as

- material inducements—money or goods

- opportunities for distinction, prestige, and personal power

- desirable physical conditions of work—clean, quiet surroundings, for example, or a private office
- pride in workmanship, service for family or others, patriotism, or religious feeling
- personal comfort and satisfaction in social relationships
- conformity to habitual practices and attitudes
- feeling of participation in large and important events (Simon, Smithburg, and Thompson, 1958: 62)

Incentives are derived from multiple sources. One source is the internal values that individuals assign to different outcomes and the activities needed to achieve those outcomes. For example, an individual with a strong preference for equitable outcomes will engage in more activities directed toward fair distribution.

A second source is the physical and technological variables that affect the transformation of activities into outcomes. Without animal or mechanical power, the amount of effort that it takes to accomplish some objectives is so great that individuals face a disincentive to attempt to achieve desired ends, such as building a permanent diversion dam. A new technology changes the relative costs and benefits so that what was once perceived as infeasible may become feasible.

A third source of incentives is the general cultural values shared by individuals in a community. Engineers, for example, are strongly motivated by professional values. The farmers using an irrigation system are motivated by ethnic, religious, caste, village, and family value systems. If the cultural values of two interacting groups differ substantially, these groups may face entirely disparate incentives even though their physical situations are relatively similar.

A fourth source of incentives is the rules-in-use that relate to specific situations in which individuals repeatedly find themselves. Rules that determine who has access rights to the water in a particular system will affect the perceived costs of various individuals who might desire to use the water. Depending on how well access rights are enforced and illegal diversions are penalized, those without access rights may consider the costs of breaking these rules sufficiently high that they do not attempt to gain access. On the other hand, where enforcement and sanctioning are not effective, those without legal access rights may pay more to divert water at night, or they may use other illegal and more expensive methods of diverting water. If the legal regulations specifying access rights are not enforced and the rules-in-use allow free-for-all access to an irrigation system, the costs of access for those with formal rights and for those with no rights may be identical.

Similarly, the rules-in-use specifying the actions that must, must not, or may be taken affect the incentives of suppliers and users in their daily activities. If farmers are supposed to rotate water to all farmers using a tertiary canal, each farmer faces a mixture of incentives when contemplating when and how much to open his field gate. Paddy rice farmers whose fields are close to the stress level face a strong incentive to open their gates immediately, whether or not their turn has come. If all farmers open their gates without coordination, however, the quantity of water that they can jointly apply to their fields is less than when a coordinated rotation system is adopted. The incentives derived from the rules-in-use have to be more powerful than the strong incentives derived from the need to keep paddy rice wet. If farmers know that they will likely be observed by a neighbor if they violate the rotation rules and that their reputations as reliable members of the community will be tarnished as a result, the costs of breaking the rules will be higher than if no social disapproval is attached to taking water when it is needed. If farmers know that everyone else is following the rotation rules and that their nonconformance might cause others to break the rules as well, the long-term negative consequences of unpredictable water availability may also dissuade farmers from an action bringing short-term benefits but threatening long-term harm.

Changes in formal regulations do not automatically become changes in rules-in-use and thus in incentives. A new regulation that greatly increases the penalty for illegally diverting water may produce entirely different changes in incentives than presumed: the threat of heavy fines may actually be used by officials to extract bribes from errant farmers as payment for ignoring infractions. Consequently, the rule-in-use may change so that diversions considered illegal by formal regulations may continue in practice as long as payments are made to the appropriate officials. Thus, the incentives facing individuals cannot be determined from a reading of promulgated laws and regulations without examining how those regulations are perceived by participants and how they fit into the physical, economic, and social context of a particular system.

Institutional Rules as Social Capital

Physical capital is the stock of material resources that can be used to produce a flow of income (Lachmann, 1978). For many engineers, an irrigation system is the equivalent of its physical capital, which consists of natural resources (rivers, springs, lakes, groundwater basins) and constructed works (headworks, canals, distribution mechanisms, field gates). But even the most modern irrigation system, complete with automatic measurement and distribution mechanisms, cannot run indefinitely without human operators. If human operators do not follow regular patterns of behavior that are expected and understood by others, especially system users, the potential flow of income from the physical capital will be severely curtailed or even eliminated. Productive patterns of behavior do not just happen.

To derive net benefits from any irrigation system, the activities of individuals must be meshed in regular and predictable patterns. In any public or private enterprise, the activities of individuals can be broadly grouped into two types: transformation and transaction (for a general discussion, see E. Ostrom, Schroeder, and Wynne, 1990). Transformation activities are directed toward changing one state of affairs into another. Transaction activities are directed toward (1) the coordination of transformation activities, (2) the provision of information, and (3) the acquisition of a strategic advantage over others.

Transformation Activities and Costs

In any large-scale irrigation project, one transformation after another must be made to bring irrigation water from a large catchment area to the farmers' fields. Figure 1 details the core flows in a canal irrigation system, as illustrated in Robert Chambers's *Managing Canal Irrigation* (1988: 36). At each of the many steps in the flow of water or goods, some kind of transformation activity is required.

FIGURE 1 Flows in Canal Irrigation Systems

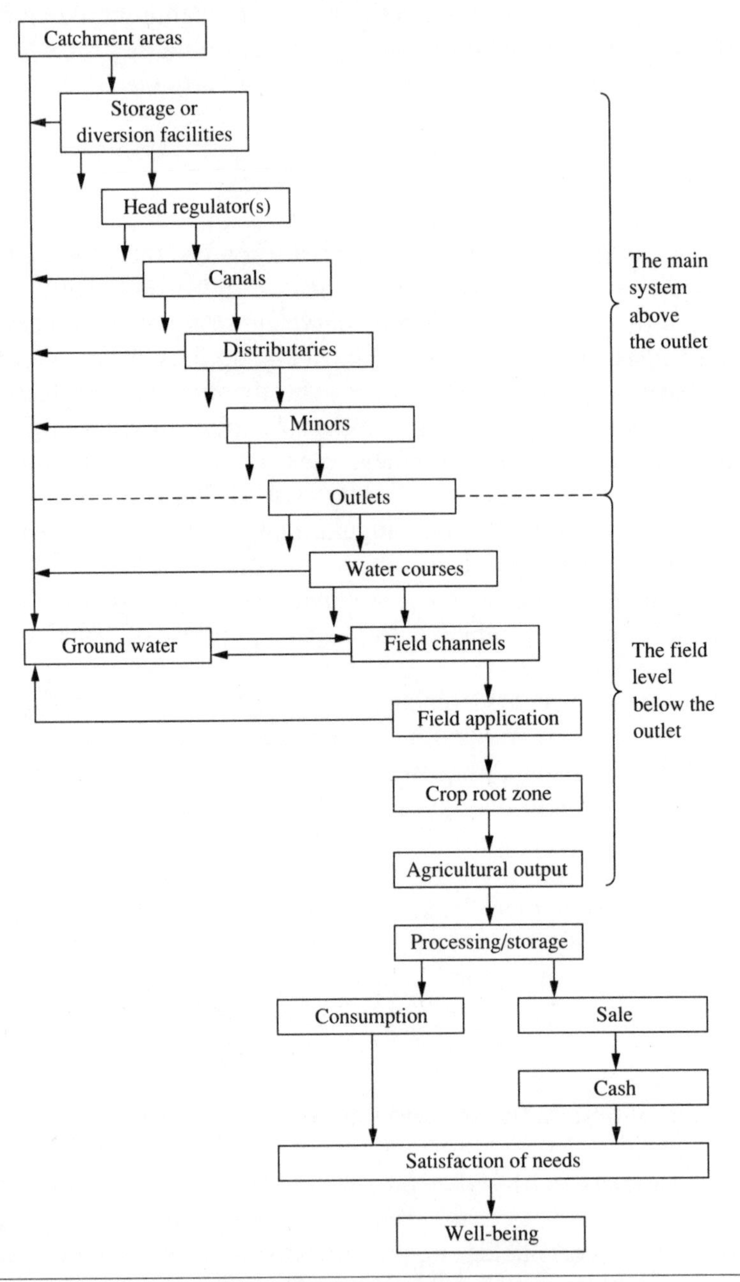

Source: Robert Chambers, *Managing Canal Irrigation: Practical Analysis from South Asia* (Cambridge: Cambridge University Press, 1988), p. 36.

How this activity is performed at each step affects what is made available at the next step and how much is wasted. Examples of transformation activities include

- diverting water from a natural water course into a constructed canal
- adjusting a barrier in a canal to raise the water level sufficiently so that it will flow into a farmer's input gate
- preparing a rice paddy to receive the season's first water
- weeding a planted field to encourage growth of a crop

When engineers compute efficiencies, they focus on transformation activities. The efficiency of an engine, for example, is the ratio of energy produced to energy used. Irrigation engineers are interested in the technical efficiency of an irrigation system in terms of the amount of water available at the farmers' intakes as a proportion of the amount of water available at the headworks. Economists are also interested in efficiency, but an economist's concept of efficiency involves the ratio of benefits to costs.

Transformation activities also involve human capital. The skill that a particular individual brings to the transformation activities he or she undertakes is a form of human capital. A single farmer working alone to enhance agricultural yield by channeling the waters of a spring located on his or her own land acquires substantial knowledge and skill over time as various combinations of crops yield more or less harvest at the end of the season. Human capital thus enables a solitary farmer to increase the productivity of investments in other inputs such as seeds, fertilizer, draft animals, or mechanical energy.

When transformation activities require the inputs of multiple individuals, good physical capital and substantial human capital are not sufficient for complex, interconnected activities to be undertaken successfully. If distributing a large flow of water without excessive waste requires that several individuals open different gates located at some distance from one another in a rapid, sequential order, the skill that each individual brings to the task of handling a single gate is not enough. Coordination is also needed. Coordination can be achieved (1) by learning how to do joint tasks better, (2) by assigning one person the responsibility to command others, or (3) by establishing a rule specifying by whom, when, and how particular activities are to be undertaken, along with establishing how that rule is monitored and enforced by participants, external enforcers, or both.

All three means of achieving coordination are forms of *social capital* (Coleman, 1986). The first form of social capital—shared learning—is a skill that those who work together acquire when they are motivated to do a good job. The other two forms of social capital are embedded in the rules jointly used by individuals. In the second form, the rules assign one person authority to command the others. In the third form, the rules specify by whom, when, and how activities are to be undertaken. All forms of social capital involve spending resources—at least time and energy—in conducting transactions with others.

Transaction Activities and Costs

Whereas transformation activities relate to changing some states of affairs into other states of affairs, transaction activities involve coordinating input activities, obtaining relevant information about transformation, or attempting to obtain disproportionate advantage from transformation activities. All transformation activities requiring inputs from multiple individuals will involve transaction activities and thus transaction costs. Coordination and information activities are essential parts of all ongoing concerns. Examples of coordination activities include

- setting the date for the first release of water from a reservoir, at which time farmers will need to be ready to make effective use of the water released
- establishing the first and last days of a budgetary cycle and the time public funds will be available for disbursement
- obtaining approval from officials and farmers concerning the design of a future project

- supervising the work of laborers digging a canal
- going to farmers' residences to collect water fees

Information activities include

- acquiring information about the hydrologic properties of various kinds of diversion works
- investigating the damage caused by a flash flood on a particular segment of a canal

Transaction activities are essential to the accomplishment of transformation activities, but the cost of transaction activities can vary dramatically depending on both the rules used and the physical environment involved. The rules that specify who is to coordinate with whom about what, and how information is to be recorded and transmitted, affect the level of transaction costs. These rules can create effective coordinating and information-sharing incentives for most participants, or they can result in frustration, delay, secrecy, and conflict rather than cooperation among individuals. The physical environment in which individuals work also affects the costs of these activities. It is more costly to communicate face-to-face in a large irrigation system than in a small system. The costs of collecting irrigation fees in a large system may be higher than in a small system.[2] In other words, the transaction costs involved in coordination and information activities can be extraordinarily high unless those who craft institutional rules find creative mechanisms to keep these costs low.

Although these costs may be high, they may be extremely difficult to measure accurately. The costs involved in coordination and information activities are rarely conceptualized or reported separately from the costs involved in transformation activities. Transformation and transaction costs are typically merged in the records of most agencies and treated simply as agency expenditures. Although some agencies obviously devote many more resources to coordination and information activities than other agencies (for a given quantity of work produced), it is difficult to obtain reliable measures of these kinds of transaction costs. It is difficult to determine, for example, the amount of time that a canal supervisor spends in actual transformation activities (opening and closing gates) versus coordination activities (scheduling work staff and opening and closing gates). The more "managerial" the position is, the more its activities are related to coordination and information and the less they are related to direct transformations.

A more difficult problem is that coordination and information activities frequently do not combine in a strictly additive fashion (Alchian and Demsetz, 1972). An effective supervisor may increase the productivity of a staff's transformation activities; thus, expenditures for effective coordination may be offset by more effective transformations. An ineffective supervisor may decrease the productivity of a staff's transformation activities; in this case, expenditures on coordination lead to even more expenditures (losses) on transformations. To add further complications, not all the coordination or information costs are contained in agency records. If users must wait many months for responses from an agency or must repeatedly provide the same information to the same agency, users also pay coordination and information costs.

The absence of coordination and information cost records does not make them any less real. Substantial amounts of time, money, and energy are spent on these activities, and the overall amount can be substantially altered by the rules-in-use and the skill of participants in transaction activities. In addition to coordination and information activities, a third class of transaction activities—and resultant costs—is potentially involved in all continuing relationships among individuals who do not share the same information, incentives, resources, and/or social norms. Such situations provide incentives for some individuals to adopt opportunistic strategies in order to obtain disproportionate benefits at the cost of others. Opportunistic behavior takes many forms. Some involve guile and deceit (Williamson, 1985). Others involve not forethought but simply actions that improve one's own situation at the cost of others. As Boss Plunkett of Tammany Hall was known to say, "I seen my opportunities and I took 'em" (Riordon, 1963).

Three types of opportunistic activities occur on many irrigation systems: free riding, rent seeking, and corruption. An example of free riding is investing time on private activities (including leisure) when others are investing in joint activities, such as canal maintenance, that increase the supply of water over time to all users. The person who shirks[3] while others work will receive a disproportionate share of benefits, because no contribution (or a reduced level of contribution) was made to the provision of benefits. The person who works while others do not feels like a "sucker" when the free riding is discovered. An example of rent-seeking behavior is trying to influence decisions made by donor agencies, national governments, or local irrigation associations about the location of and subsidies to irrigation facilities. The person who seeks rents receives a disproportionate profit on private activities because the value of his or her assets is artificially increased. An example of corruption is withholding the delivery of water to those entitled to it in order to receive illegal side-payments of money, commodities, or special favors. The person who engages in corruption receives a disproportionate gain by using his or her power over the allocation of valued resources to extract an illegal payment from someone else.

Although free riding and corruption are relatively well understood, noneconomists (and even some economists) often seriously misunderstand the terms "rent" and "rent seeking." Because the creation of rents and the seeking of rents are so important to an understanding of the perverse incentives related to irrigation institutions, it is important to clarify these concepts.

Rents are the excess profits earned by a holder of a property right that exceed what could be obtained in a competitive market. "They can be created purposefully; monopoly rents, for example, accrue to those who restrict competition in product markets" (Bates, 1987: 35). Individuals also may obtain rents because they are just fortunate enough to own rights to property with special advantages, such as fertile fields or an area with mineral deposits. The possibility of deriving rents generates incentives for some to seize control over rent-generating properties, to invest in activities to secure subsidies from others, or to exclude potential competitors. These activities devoted to enhancing rents are called *rent seeking* (Krueger, 1974; Tollison, 1982; Buchanan, Tollison, and Tullock, 1980).

Edward Vander Velde (1980) paints a vivid picture of how a new irrigation project in rural India, served by the Dhabi Minor canal system (a part of Bhakra-Nangal project), increased the value of the land owned near the project and strengthened the already substantial economic, social, and political power of members of a higher social caste. The value of irrigable land rapidly approached twice the value of dry cropland. Most of the land in the area was owned by higher-caste farmers. Sharecropping leases made with poor lower-caste farmers were generally of the most exploitative nature. One-third of the production was kept by the cultivator, and two-thirds was turned over to the landowner—an illegal but nonetheless frequently practiced tenurial arrangement (Vander Velde, 1980: 319–21). The formula devised by the state irrigation agency to determine how much water each farmer was to receive and the way the system operated in practice also gave the richest farmers access to the most water. As Vander Velde (1980: 324–27) indicates:

> irrigation development and the methods of operation of the irrigation system transformed these large holdings, now comprising mixed amounts of highly valuable irrigable land and much less desirable dry crop land, into an even greater asset than they had been. Because the length of farmers' irrigation turns and thus the amount of water to which they are entitled are determined by the size of the cultivation unit in the command of the system, there is even greater reason to retain title to the largest area possible because by doing so one maximizes access to the most scarce resource in this environment.

This is a description of how rents are created by new irrigation systems. It is no wonder that rich farmers spend time and effort trying to influence politicians to bring irrigation projects to their area. Nor is it any wonder that politicians recognize that the favors they extend to those who support projects or subsidies in general are a method of acquiring additional political influence.[4] Tragically, the vast opportunities for economic and political gain that large-scale river-basin developments have created have also led in some cases to exacerbated ethnic or religious conflicts and even increased bloodshed.[5]

All opportunistic activities produce short-term costs for others and, potentially, long-term costs for everyone involved. In the short term, the person engaged in opportunistic behavior shifts costs to others. If opportunistic behavior is considered likely, individuals may prepare for the worst by adopting cautious strategies to protect against exploitation (Scharpf, 1990). When all individuals are cautious and protective, however, they may miss many opportunities for mutually productive gains. Thus, the major costs of opportunistic behavior are the many productive activities that are not undertaken because institutional arrangements and social norms have not been developed to protect individuals against opportunism. Shifted costs and forgone opportunities are real costs. These real costs may not be recorded, however, in any regular fashion. Hence, they are even more difficult to measure than information and coordination costs.

Opportunistic activities are infrequently discussed in treatises on irrigation or on development processes more generally. Some scholars and practitioners wish to describe the world without including the human capacity for avarice and for taking advantage of others. These activities are discussed at length in this study because of the potential for substantial losses resulting from opportunism, *not* because it is assumed that all individuals are opportunistic all the time. Many public officials do not ask for or accept bribes even when surrounded by colleagues who engage openly in corrupt practices; many individuals are willing to contribute to the provision of joint goods even when only a few others join them in these activities; and many powerful individuals do not try to influence public policies so that the land they own will balloon in value or the prices they pay for inputs will be artificially low.

But for all the individuals who refrain (most of the time) from opportunistic actions, others will avidly adopt opportunistic strategies at the slightest temptation. The organization of irrigation institutions in much of the developing world unfortunately creates many opportunities for free riding, rent seeking, and corruption. The costs of providing irrigation water are much higher in many settings because of the prevalence of these activities, and the distribution of irrigation benefits is frequently distorted.

When institutions are well crafted, opportunism is substantially reduced. The temptations involved in free riding, rent seeking, and corruption can never be totally eliminated, but institutions can be devised to hold these activities in check. In order to decrease opportunistic behavior, coordination activities, such as monitoring and sanctioning, may have to be increased. The costs of monitoring and sanctioning activities to eliminate *all* instances of opportunistic behavior would be excessive. Controlling opportunistic behaviors must involve keeping the temptations to engage in these activities low and the likelihood of discovery high.

The full range of transaction costs involved in exchange and production activities has only recently been considered by scholars and practitioners interested in the effects of using different institutional arrangements for accomplishing diverse tasks. The models used by neoclassical economists to describe exchange behavior in markets most frequently assume away all transaction costs and presume that ignoring the "friction" associated with transaction activities does not detract from the power and usefulness of their models. In markets where the assets and products involved are homogeneous and where large numbers of individuals interact, transaction costs can be ignored without great loss to the usefulness of findings. Many markets, however, involve asset specificity and small numbers (Williamson, 1979, 1985). In these settings, ignoring transaction costs yields theoretical explanations and predictions that are not supported by empirical evidence (see North, 1989). The importance of costs that result from a lack of information and from the opportunistic behavior of participants has received a growing recognition in the work of scholars who associate themselves with the "new institutional economics."[6] The major accomplishment of scholars working in this tradition has been the demonstration of the strong influence of diverse institutions in counteracting different types of opportunistic behavior and affecting the costs of obtaining accurate time and place information.

Until recently, administrative theorists have largely ignored transaction costs other than those associated with coordination activities and the acquisition of technical or scientific information. For example, the amount of attention that Robert Chambers devotes to the problems associated with corruption in *Managing Canal Irrigation: Practical Analysis from South Asia* (1988) is at odds with most treatments of management problems in general and irrigation in particular. His subtitle reflects his concern for analyzing many aspects of running irrigation canals that are not contained in more theoretical treatises. Chambers's book is refreshing, given his frank assessment of many "practical" problems. For his discussion of corruption, he and others interested in this problem are deeply indebted to the pioneering work of Robert Wade (1982a, 1982b, 1985). Recent work from an institutional perspective has demonstrated that the specific rules used to coordinate activities within and among administrative agencies strongly affect the level and type of transaction costs involved (Hechter, 1987; Breton and Wintrobe, 1981).

The institutional capital present in any particular set of suppliers and users may enable these individuals to cope effectively with both transformation and transaction costs and thereby achieve amazing levels of productivity with only primitive forms of physical capital. The *zanjera* institutions of the Northern Philippines (Siy, 1982), the *Subaks* of Balinesia (Geertz, 1980), and many of the farmer-managed systems of Nepal (Pradhan, 1989a) are all remarkable for the high levels of effectiveness achieved from systems whose physical capital appears outdated to many contemporary engineers. The complex network of relationships established between government officials, farmer representatives, and the farmers themselves on many irrigation systems in Taiwan (Levine, 1980; Bottrall, 1981; Moore, 1989) illustrates that it is possible for effective social capital to be crafted on irrigation systems constructed, owned, and "operated" by a national irrigation bureaucracy. The remarkable improvements achieved as a result of a program to strengthen farmer organizations on National Irrigation Agency systems in the Philippines illustrate the possibility of learning from experience to improve jointly managed systems (Korten and Siy, 1988). The Gal Oya experience in Sri Lanka, in which institutional catalysts worked with farmers to learn about their problems and help them build a nested set of organizations from the ground up (Uphoff, 1985), is similarly revealing.

Yet the institutional capital present on many irrigation systems constructed during the past three decades in developing countries is often sadly lacking. William Ascher and Robert Healy (1990) document the lack of investment in institutional arrangements in two major irrigation projects in India (the Jamuna project in Assam and the Nalganga project in Maharashtra). In both cases, planning focused entirely on the construction of major physical works and presumed that the farmers would automatically organize to construct, operate, and maintain field channels to get water from the system to their fields. Construction of the Jamuna project was completed in May 1969, costing approximately $8.8 million (Ascher and Healy, 1990: 147). Five years later, less than a third of the planned service area was receiving irrigation water. An *ex post* evaluation discovered that the root of the problem was the refusal of the farmers to construct field canals.

> The disastrous oversight was engendered by the project initiation approach of the experts and authorities concerned. . . . The farmers had the time and physical resources to construct the channels.

> Yet the channels were slow to come. . . . The obvious reason for this, which the project authorities did not anticipate and failed to learn because the beneficiaries were not involved in project design and implementation . . . ,was that the farmer closer to the headwaters had no incentive to devote his own (or hired) labor to constructing channels that would conduct the water through his own field into another's. (Ascher and Healy, 1990: 148–49)

In other words, a project whose physical works cost close to $9 million was producing a small proportion of its projected benefits as a result of a lack of investment in crafting institutional arrangements among farmers to construct (and eventually operate and maintain) the simplest type of water conveyance channels. Social capital is not automatically or spontaneously produced.[7] It must be crafted.

Notes

1. Common knowledge is an important assumption that is frequently used in game theory and is essential for most analyses of equilibrium (Aumann, 1976).

2. Thus both the size of system and the specific rules affect transaction costs. Both elements are reflected in the estimates made for collecting irrigation fees in Egypt, which vary from a low of under $1 to over $7 per acre depending on the type of water fee assessed (Easter, 1985: 16).

3. Shirking is the term used most frequently to refer to free riding on the job. A water-gate operator who stays in a nice dry office during the monsoon season rather than doing his assigned work is shirking. The operator is paid but does not do the work that is supposed to be done.

4. See Craven et al. (1989, Vol. III: A29) for a description of the "land rush" in Somalia in anticipation of the construction of a dam on the Jubba River. Large tracts of land have been registered by external investors and speculators, some of whom were civil servants.

5. See Scudder (1990) for a discussion of genocide and civil wars associated with large-scale river-basin developments in Mauritania, Somalia, Sudan, and Sri Lanka.

6. For a review of this literature as it relates to development issues, see the special issue of *World Development* (vol. 17, no. 9, 1989), edited by Irma Adelman and Erik Thorbecke, on "The Role of Institutions in Economic Development."

7. The term "spontaneous order" is frequently used to describe a wide diversity of patterns of human order. These patterns share one

characteristic—they were not designed by a central governmental official. They differ on many other dimensions. A path through a wooded area may well be the result of many different individuals spontaneously choosing to follow a deer trail or the trails of other humans. But using the term "spontaneous" to describe the coordinated activities of farmers to build, operate, and maintain field channels overlooks the substantial amount of time these farmers invest in working out acceptable rules and monitoring conformance to these rules. Use of the term "spontaneous" by academics fosters the impression that these efforts will automatically spring forth.

CHAPTER THREE

Crafting Institutions

The term "crafting" with reference to the development of institutions emphasizes

1. the artisanship involved in the design, operation, appraisal, and modification of rule-ordered behavior (V. Ostrom, 1980)

2. the ongoing nature of "getting the process right" (Uphoff, 1986)

Crafting institutions for the supply and use of irrigation systems is challenging and requires skill in understanding how rules, combined with particular physical, economic, and cultural environments, produce incentives and outcomes. There is no "one best way" to organize irrigation activities (Coward, 1979; Chambers, 1980; Levine, 1980; Uphoff, 1986; E. Ostrom, 1990). Rules governing the supply and use of any particular physical system must be devised, tried, modified, and tried again, and considerable time and resources will be invested in learning more about how various institutional rules affect participants' behavior. Thus, the choice of institutions is not a "one-shot" decision in a known environment but rather an ongoing investment in an uncertain environment.

Crafting Institutions as an Investment Process

Devising, testing, revising, monitoring, and enforcing a set of working rules to structure irrigation activities is a time-consuming endeavor. The time *invested* in constructing and operating a better institutional structure is similar to the time invested in building and operating a better physical structure. It results in *shared knowledge* about how to coordinate the inputs of many individuals in a series of complex, interdependent, and time-dependent activities. Viewing the design, trial, modification, and monitoring of institutions as an investment process has several immediate implications. To invest in *any* capital structure, whether it be physical or institutional in form, requires that the time and effort which would otherwise be allocated to obtaining immediate benefits (including leisure) be diverted instead to activities that will achieve an uncertain flow of benefits over a long time-horizon. Those who heavily discount future returns will not make such investments. Those with short time-horizons will attempt to do as well as they can within the constraints of available physical capital (the irrigation works) and social capital (the rules-in-use and shared skills of the suppliers and users of the irrigation works).

Farmers who are on the verge of dire poverty cannot afford to divert many resources from activities directly related to short-term benefits in return for uncertain long-term benefits. If they cannot feed their families and pay for their land, they will not be around to reap the long-term benefits of investing in either new physical improvements or new ways of coordinating their activities with others. Similarly, public officials who do not expect to be assigned to the same location for more than a few years have less motivation to invest time and effort in improving capital structures in that location than those who have a long-term commitment.

Many irrigation systems that have been constructed in developing countries since the 1950s involve both users and suppliers who have relatively short time-horizons; their actions, however, have long-term effects on both social and physical capital. On large irrigation settlements, for example, eligibility criteria have frequently required a settler to be landless and to have a large family (Harriss, 1984: 325). Recruitment using these criteria yields a heterogeneous set of individuals coming from different regions, kinship groupings, and ethnic and religious backgrounds, many of whom have very limited individual capital. *No* social capital exists when large numbers of heterogeneous individuals are placed in a strange terrain. With few acquired farming skills and with large families to feed (by project requirement), the initial settlers are challenged just to make ends meet and keep the land they were assigned. Many do not succeed. Eventually, some sell their land and return to the ranks of the landless.

Settlement rules sometimes require that land allotments distributed to new settlers be inherited intact. Although the attempt to avoid extreme fragmentation of land holdings is understandable, the unfortunate result is a proliferation of sibling rivalries and a tendency for young men to seek opportunities elsewhere. On some projects, the proportion of young men remaining to work on the family farm has fallen as low as 10 to 15 percent (Harriss, 1984: 328). In such situations, neither parents nor offspring develop the long time-horizon needed to change institutional rules and increase long-term net benefits.

In many countries, the staff who collect irrigation fees for particular projects or administrative districts are frequently engaged in a "transfer trade," meaning they will stay in one position for no longer than two or three years. Most national agencies routinely rotate officials from one post to another. The presumption underlying this policy is that rotations curtail corruption and favoritism. However, as has been documented most thoroughly in India, this result does not always occur. Sharan and Narayanan (1983) found that in Banowara and Dungapur districts, collectors averaged only fourteen months per assignment. Between 1948 and 1981, the longest stay in this position in either district was under three years. Where politicians control postings, as they have in India, transfers become "a powerful instrument for punishment and patronage" (Chambers, 1988: 185). Irrigation posts are auctioned off by politicians to competing engineers.

> Posts were known by their nominal prices—a "one lakh post," a "five lakh post" but additional payments might be demanded during the normal two-year tenure, particularly if there was an election. To remain beyond the two years required a further payment. Moreover, security in post even for the understood two years was far from assured.... Astonishingly [superintending engineers] could pay 40 times or more their annual salary. (Chambers, 1988: 186)

Such a system offers two powerful incentives against investing in improvements to irrigation system operation. First, the short tenure reduces the officials' time-horizons. Second, officials have had to pay such a high price for their postings that considerable effort must be devoted to gaining illegal income from contractors (through kickbacks and payments to ignore shoddy work) and farmers (through payments for water delivered or a lack of enforcement of formal regulations). Thus, if system operations were improved, the income that an engineer could obtain from a posting might actually be reduced.[1]

On settlement projects where agency personnel face uncertain futures, *no one* has the requisite time-horizon to invest in social capital. Investments in physical capital may be shoddy and purposely below standards. Project planners who presume that spontaneous organization will emerge have not thoroughly analyzed what is involved in building social capital. Evidence indicates that the motivation to invest in social capital exists on established irrigation projects where (1) farmers have long time-horizons, (2) they face sufficient scarcity that they are motivated to invest in organizing themselves, and (3) they are assured that organization could make a substantial difference in their yields (Wade, 1988; Uphoff, Wickramasinghe, and Wijayaratna, 1990).

Multiple Layers of Rules-in-Use

When investments are involved, two levels of analysis are required. First, an analyst needs to understand what is happening at an operational level, where individuals attempt to do as well as they can *within* the physical and institutional constraints as they exist. Second, an analyst needs to consider what options are available to change those constraints. Considering these changes is like calling a time-out during a game to reconsider the rules of the game itself. This type of shift is involved when farmers consider new technologies on their farm or the suppliers of an irrigation project consider installing a new type of control gate (Nelson and Winter, 1982; Dosi, 1988).

Initial rules are nested within another set of rules that define how the initial set can be changed.[2] This nesting of rules is similar to the nesting of computer languages. What can be done at one level depends on the capabilities and limits of the software (rules) at that level as well as the software (rules) at a deeper level and the hardware (the physical works). When considering *institutional change,* as contrasted to action within institutional constraints, it is essential to recognize two factors:

1. Changes in the rules used to order action at one layer occur within a currently "fixed" set of rules at a deeper layer.

2. Changes in deeper rules are usually more difficult and more costly to accomplish.

It is useful to distinguish three layers of rules that cumulatively affect irrigation systems (Kiser and E. Ostrom, 1982). *Operational rules* directly affect the day-to-day decisions made by users and suppliers concerning when, where, and how to withdraw water; who should monitor the actions of others and how; what information must be exchanged or withheld; and what rewards or sanctions will be assigned to different combinations of actions and outcomes. *Collective-choice rules*, which indirectly affect operational rules, are used by irrigators, their officials, or external authorities in making management policies. A change in policy implies a change in operational rules. *Constitutional-choice rules* determine (1) who is eligible to participate in the system and (2) what specific rules will be used to craft the set of collective-choice rules, which in turn affect the set of operational rules (V. Ostrom, 1982).[3]

The linkages among these rules and the related arenas in which individuals make choices and take actions are shown in Figure 2. The processes of allocating water, clearing canals, and monitoring and sanctioning the actions of irrigators and officials occur at the operational level. Policy making, management, and policy adjudication occur at the collective-choice level. Formulation, governance, adjudication, and modification of constitutional decisions occur at the constitutional level.[4]

Rules are changed less frequently than the strategies individuals adopt within rules. Changing rules at any level increases the uncertainty that individuals face in making strategic choices. Rules provide stability of expectations, and efforts to change rules rapidly reduce that stability. It is usually the case that operational rules are easier and less costly to change than collective-choice

FIGURE 2 Linkages among Rules and Levels of Analysis

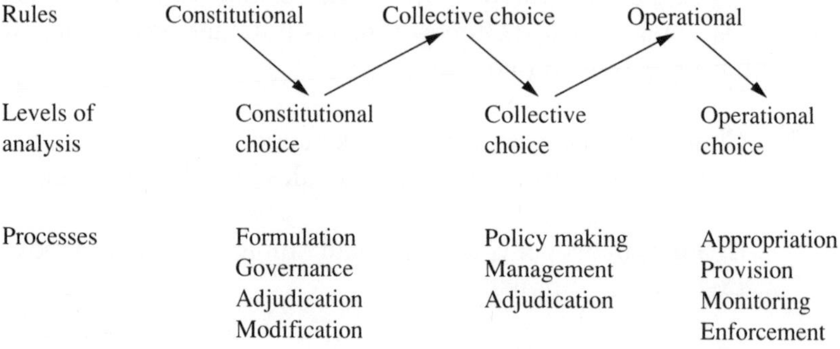

Rules	Constitutional →	Collective choice →	Operational →
Levels of analysis	Constitutional choice	Collective choice	Operational choice
Processes	Formulation Governance Adjudication Modification	Policy making Management Adjudication	Appropriation Provision Monitoring Enforcement

Source: Elinor Ostrom, *Governing the Commons: The Evolution of Institutions for Collective Action* (New York: Cambridge University Press, 1990), p. 53.

rules, which in turn are easier to change than constitutional-choice rules. If constitutional-choice rules *can* be changed easily, preemptive decisions at that level may induce serious instabilities at the collective- and operational-choice levels. Rapid changes at a constitutional level will seriously erode the mutual expectations about how future collective-choice decisions will be made, which in turn will affect operational-level decisions.

The results of changing deeper layers of rules are more difficult for participants and scholars to analyze. Deciding whether the constitution of an irrigation association should establish a legislative body of five or nine members depends upon the physical characteristics of a system and the governance systems that the participants are accustomed to using.[5] A change in this constitutional rule usually will not make an immediate and noticeable difference. Change at the constitutional level is reflected in a change in the pattern of collective-choice decisions because these constrain or open up possibilities at an operational level.

Multiple Sources of Rules-in-Use

At each level of analysis, there may be one or more decision-making arenas. An arena is simply the setting in which a particular type of action occurs; arenas include such formal settings as legislatures and courts, but they can also include informal settings, such as places where people regularly gather to talk with one another. Decisions about the rules that will be used to regulate operational-level choices are made in one or more collective-choice arenas. When irrigators want to change some of the collective-choice rules concerning appropriation and provision, they may meet in a local coffeehouse, schedule a co-op meeting, or form an organization—such as a water user association—specifically for the purpose of managing and governing the system. If the irrigators or project officials, working together or independently, cannot change at least some of the operational rules, the only arenas for collective choice are external to a particular system. In such cases, rules are written by external administrative agencies, elected representatives in local or national legislatures, or judges in judicial arenas. Such rules will rarely reflect the particular circumstances facing users and suppliers on a particular system.

A single arena rarely corresponds exclusively with a single set of rules. Most frequently, several collective-choice arenas affect the set of operational rules. Decisions made in national legislatures, ministries, and courts about the practices to be followed by particular types of irrigation systems—if these practices are given legitimacy in a local setting and enforced—are likely to affect the actual operational rules-in-use. Similarly, formal and informal constitutional-choice processes may occur in local, regional, national, and/or international arenas. The relationships between formal and informal collective-choice arenas and resulting operational rules are illustrated in Figure 3.

That working rules may have multiple sources and include *de facto* as well as *de jure* rules greatly complicates the problem of understanding what is happening in particular irrigation systems. As discussed previously, the absence of national laws regulating the property rights to water or responsibilities for system maintenance is not equivalent to the absence of effective rules for a particular system. Local users and suppliers may have invested in the development of working rules over a long period of time. Such rules may or may not lead to efficient and fair management of a system, but they do affect the strategies that users and suppliers perceive to be available, the actions they take, and the consequences that follow.

FIGURE 3 Relationship between Formal and Informal Collective-Choice Arenas and Operational Rules-in-Use

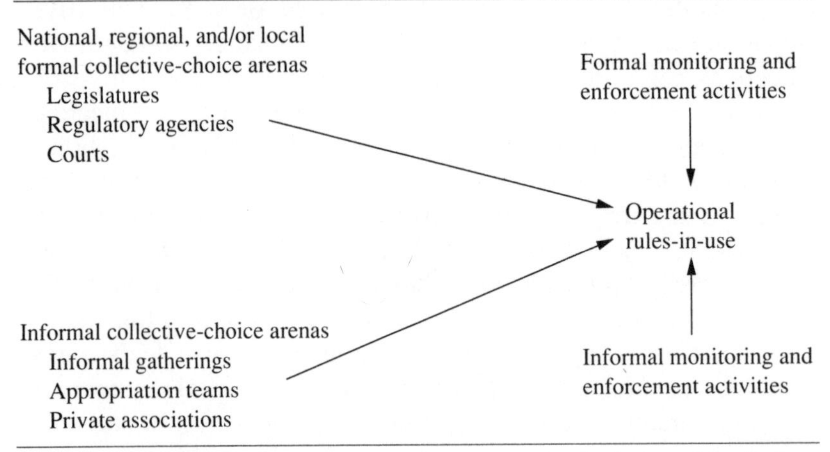

Source: Elinor Ostrom, *Governing the Commons: The Evolution of Institutions for Collective Action* (New York: Cambridge University Press, 1990), p. 53.

Crafting Rules for Varying Environmental Conditions

If local users and suppliers participate in crafting at least some of the rules affecting their operational choices, system performance is more likely to be enhanced. One reason for this is the vast variety of environmental conditions that affect the physical operation of any particular system. "Each canal irrigation system has a distinct constellation of many variable parts" (Chambers, 1988: 211). Efforts to classify systems for the purpose of devising standard rules for use on all systems in a particular category have not proved useful, nor will they. Analysts have attempted to classify irrigation systems by such variables as

- size
- type of water source
- soil type
- crops irrigated

- physical topography
- climate

As Chambers points out, however, these simple "classifications cross-cut each other. They also omit many vital aspects" (1988: 211). Among the omitted aspects, Chambers lists the following important variables:

- water adequacy and quality of delivery
- canal capacity in relation to peak demand
- physical capacity to control flows
- rights to water
- financial responsibilities
- political organization and environment
- farm sizes
- farmers' relations and communications with staff
- labor availability

In addition to the sheer number of physical characteristics that affect the day-to-day problems faced in operating an irrigation system, the specific configuration of variables in any irrigation system is usually more important than any single variable. A large physical system with many smaller storage facilities is quite different than a large system without any storage facilities below the intake. Given the large number of variables, the number of configurations of variables is immense, and no standard set of rules for an entire region can possibly work well.

Multiple-purpose systems that involve both in-the-channel and on-the-land uses of water are even more complex. The large-scale dams used for both irrigation and flood control involve operational procedures unlike those used for irrigation alone. An empty reservoir is preferred for flood control, but a full reservoir is preferred for irrigation. Devising operating rules for providing irrigation while simultaneously trying to prevent damaging floods is substantially different from using a system for one purpose only.

Furthermore, operational problems may differ from season to season. A set of rules devised on the basis of specific system characteristics may work well during a monsoon season when water is allowed to flow freely, but it will not work well during a dry season when water is scarce and must be allocated carefully. Most irrigation systems where suppliers and users have crafted at least some of the key rules-in-use have more than one allocational rule, depending on the availability of water. These rules can vary dramatically in many systems from season to season and from year to year.

In the long-enduring irrigation institutions for managing *huertas* in southeastern Spain, for example, local officials determine the basic rules for allocating water in response to three environmental conditions: abundance, seasonal low, and extraordinary drought (Maass and Anderson, 1986). A tight rotation system is used when seasonal-low water conditions are present; this is the most frequently observed condition. In rare times of abundance, water is allowed to flow in all canals, and farmers can take as much water as they want, whenever they want. When an extraordinary drought is declared, an administrative official takes direct charge of allocations and attempts to send water to the driest fields. Barker et al. (1984: 38–39) describe a traditional system in Taiwan (Yun Lin), in which traditional property rights assignments give the farmers on some canals considerably more water than others during times of abundance. When water is scarce, however, these farmers switch to a larger system with improved conveyance structures and maintenance. As part of the agreement to be included in this larger system, the set of traditional property rights is replaced by a "technical" set that distributes water equally to various parts of the system. The switch to the second set of rules is made in small irrigation association meetings when irrigators collectively agree that the water supply is low. In many Asian irrigation systems where paddy rice is a major crop grown, water is distributed continuously during the monsoon rainy season but rotated during the drier seasons.[6]

Whether a system is capable of storing water in a reservoir or can augment surface water with groundwater makes a substantial difference in the predictability of water deliveries, the institutional arrangements that are possible, and the feasibility of market arrangements for buying and selling water. Before a farmer purchases water, he or she needs assurance that water purchased will actually be available. No such assurance can be given in systems without at least some minimal storage capacity. The only Spanish *huerta* to develop a system for auctioning water, for example, is in Alicante, where the Tibi Dam was constructed in 1594. Farmers can receive information about the quantity of water that is stored in the dam and available for release during rotation periods (Maass and Anderson, 1986). Consequently, they are assured that the water they purchase will actually be available. In India, extensive markets for water have also evolved where farmers are able to purchase defined quantities of groundwater from owners of deep-well turbine pumps (Shah, 1985, 1986). In southern California, sophisticated management institutions, including an active market for groundwater rights, are built on the foundation of negotiated court settlements that define specific rights to groundwater (see E. Ostrom, 1990; Blomquist, 1992).

Environmental variability also affects the challenges faced in maintaining an irrigation system. In a hilly region that is periodically pelted with torrential rains, maintaining diversion works and/or canals requires constant diligence and immense investments in labor and materials. A small break in a canal that appears early in the morning after a heavy rain may become a gaping hole by mid-afternoon if not discovered and repaired immediately.

In addition to the changes wrought over time by climatic conditions, dynamic processes at work in the external environment of many irrigation systems can be the source of major problems in crafting institutions. Rapid changes in the *relative* values of such diverse factors as market prices for labor, agricultural inputs, or commodities are particularly challenging. It is difficult to adjust locally devised rules rapidly enough to counteract price changes without undercutting the stability of expectations. A set of rules devised for one set of relationships between the value of land and water may not perform well when relative values shift dramatically.

Important environmental differences between irrigation systems (and even on the same system during different parts of the year) are not taken into account when national or regional governments attempt to specify the rules to be used on all systems within its jurisdictions. Each of the states of India, for example, attempts to specify the same water-allocation rules throughout its domain regardless of differing hydrologic or meteorologic conditions (Bottrall, 1981).

Crafting Rules Related to Varying Cultural Traditions

Although the climate, geology, soil conditions, terrain, and physical works of an irrigation system are obvious constraints, the shared belief systems of a particular region, caste, religion, or ethnic group also need to be considered in institutional design. When shared understandings exist concerning the fairness of diverse allocation rules, appropriate leadership positions, and the rights and duties that individuals possess in relationships with one another, the basic repertoire of rules that can easily be used by suppliers and users of an irrigation system is circumscribed. Some rules that would seem to be more efficient or fairer to an outside observer may not be included in this basic set of rules. If external authorities attempt to impose rules outside this set on unwilling recipients, it is unlikely that such rules will be followed.

The rules used in a cultural tradition are forms of shared knowledge. Farmers who have used a particular leadership selection mechanism for other purposes have an initial understanding—and basis for evaluation—of the likely consequences of using a similar device for selecting leaders of an irrigation organization. Labor-sharing formulas used successfully to mobilize adequate numbers of able-bodied workers for analogous purposes may be used to accomplish a different task. Since investing in new rules is always risky, it is not surprising that investors are more willing to work with rules whose outcomes they have witnessed than with rules whose outcomes are uncertain.

Crafting Rules to Counteract Opportunistic Behavior

Reducing the level of opportunistic behavior is a major problem for all irrigation systems. Many of the shared conceptions and norms of behavior that are collectively referred to as "culture" have evolved as a form of social capital to counteract opportunistic behavior. If participants do not view the specific rules crafted to organize a particular irrigation system as being appropriate, behavior that violates accepted norms of behavior may not be sanctioned. If formal structure is viewed as illegitimate, behavior that undercuts the maintenance of that structure will not be viewed with disapprobation.

Consequently, when central agencies attempt to impose standard organizational rules on all irrigation systems within a large jurisdiction, these rules may fail for several reasons: (1) the standard set of rules may not adequately cope with the configuration of physical variables that characterizes a specific system; (2) the rules may be "foreign" to local participants, who are uncertain of their consequences or how to implement them; and (3) other aspects of social capital—in particular, the norms of behavior used to counteract opportunistic behavior—may not be mobilized, because the "foreign" organization is not viewed as legitimate.

As discussed in Chapter 2, opportunism can take many forms. Free riding, rent seeking, and corruption are the three forms of opportunism that are the most prevalent in irrigation systems. In any situation where farmers do not contribute to the maintenance of a system, the difficulty of preventing them from benefiting from the construction, repair, or maintenance activities performed by others creates the potential for free-riding behavior. Obtaining control over resources to make a greater profit than would be possible under competitive circumstances—rent seeking—can occur anywhere (see Repetto, 1986). Soliciting illegal side payments in exchange for favors—corruption—is also a widespread threat to efficient and fair operations in all settings.

If free riding becomes the dominant mode of behavior in irrigation systems—which is certainly possible—all users are ultimately hurt. Without resource inputs in the form of fees, labor, or materials, a system cannot be repaired and maintained for long. When canals fill with silt, sufficient water does not flow through them to supply tail-end farmers. If farmers are assured that benefits exceed costs, that their inputs are necessary, and that most farmers will participate, they will frequently forgo free riding and contribute substantial amounts of labor. In other words, farmers want to be protected against being the "suckers" who participate while free riders devote themselves to private activities and snicker at the gullibility of those who do participate.[7]

Free riding involves passive behavior—free-riding farmers let others contribute while they refrain from contributing to the provision of a collective benefit. Rent seeking, on the other hand, involves active efforts to obtain disproportionate advantage from profit-making activities.

> Potential recipients of economic rent compete for them, not by outbidding rivals in the marketplace through superior economic efficiency and foresight, but by trying to control the people who allocate them. Political manipulation, intimidation, and corruption replace economic efficiency as ways to get ahead. Inevitably, most of the available rents are captured by those with power, influence and wealth, and rent-seekers think that using the resource efficiently is much less important than gaining control of the allocation mechanism. (Repetto, 1986: 14)

Once rent seekers have gained special privileges, they can use the substantial profits they gain to preserve and expand their excessive gains.

Rice farmers and influential politicians have much to gain by seeking external funding from donor agencies and by continuing to use fiscal systems that assess the general taxpayer, rather than the irrigators, for the cost of operating and maintaining large-scale irrigation systems. Institutional rules that require irrigators to cover the cost of operating and maintaining their systems (and to contribute to the recovery of the initial investment) can help curb rent-seeking behavior. Nationwide directives that charge farmers for the water they use may be completely ineffective unless an agency is willing to devote substantial resources to monitoring and sanctioning noncompliers. Farmers are actually willing to pay considerably more money than the nominal fees written into most national legislation. But this willingness to pay for water they are assured of obtaining may also be accompanied by a willingness to buy directly from a deep-well pump owner or to pay a bribe in return for assured delivery. Rent seeking cannot be curbed by legislative fiat alone without real efforts to increase system performance so farmers perceive definite benefits from paying water-user fees. Since fees frequently are not part of the revenue used in operating a project (i.e., fees are deposited in the general treasury), it is hard to relate increased fee collection to improved system performance.

Devising institutions that do not allow public officials full control over essential resources can help to reduce corruption. On those self-organized irrigation systems where corruption is typically low, the resources needed to produce jointly beneficial outcomes are rarely transferred to or controlled by officials. Many of the resources mobilized to operate and maintain such systems are in the form of labor. Since users know exactly where their labor is being allocated on work days, they can insist that their work be entirely devoted to the upkeep of the system, rather than to improving an official's personal property. Once input resources are mobilized in the form of money rather than labor, careful record-keeping that is open for public inspection is a critical requirement for circumventing corruption.

Crafting Monitoring, Sanctioning, and Conflict Resolution Mechanisms

It is as important to devise workable procedures for monitoring the behavior of irrigation water suppliers and users, sanctioning nonconforming behavior, and resolving conflict as it is to devise the rules themselves. Where substantial temptation exists to engage in opportunistic behavior, no set of rules will be self-enforcing (V. Ostrom, 1980). Whether the behavior of participants conforms to the rules-in-use must be determined by those involved and, potentially, by officials and/or external guards. Those who do not conform to these rules need to have sanctions imposed upon them. As soon as some individuals monitor others and impose sanctions, conflict will occur over rule interpretation, the facts of the event being sanctioned, and the appropriate level and type of punishment.[8] Lack of monitoring, sanctioning, and fair, inexpensive arrangements for conflict resolution can all undermine a complex system of mutual expectations and commitments.

Michael Hechter (1987: 150–57) identifies several strategies that groups can adopt to increase the effectiveness of monitoring, including (1) increasing visibility through architecture and the creation of public rituals and (2) minimizing errors of interpretation by establishing clear-cut rules and recruiting participants who share similar views. The physical design of an irrigation system and the devices and rules used by farmers in distributing water can affect how costly it is to monitor and how likely it is that rule-breaking behavior will be detected. Systems that are constructed so that the actions of farmers taking water are visible at low cost to others awaiting their turns increase the prospect of effective monitoring. Similarly, rules requiring a sequential rotation system along any one canal greatly reduce the ambiguities of who is supposed to be taking water and who is next in line. Furthermore, such rules bring those who are most directly affected to a similar physical location at overlapping intervals of time.

Sequential rotation, which is frequently used in farmer-managed systems, is criticized by irrigation engineers as being too rigid and technically inefficient. If a farmer has a higher-value use for available water but is not next in line, it is difficult to adjust these sequential water distribution systems to deliver water to whoever will receive the highest value from it. There may be other factors to consider in evaluating the allocation rules of an irrigation system besides the short-run efficiency of water use. If farmers cannot effectively monitor an allocation scheme at a relatively low cost, short-term efficiencies can rapidly be lost as monitoring declines and improper allocations (theft) rise (Weissing and E. Ostrom, 1991). Farmer-constructed irrigation systems are frequently divided into many discrete physical units within a larger system. At times, they are "arranged so that each unit is served directly from the main canal or a lateral and is not dependent on a water supply that passes over the territory of another mini-unit" (Coward, 1980: 207). This type of physical design has two consequences. First, the number of farmers whose actions directly affect one another is small, even when the number of farmers served by the entire system is quite large. Second, the efficacy with which each farmer can monitor other farmers is also relatively high.

Of course, if a large system is to be divided effectively into relatively separable subunits, clear rules for allocating water must exist and be monitored. Farmer-owned "federated" systems tend to organize themselves around mini-units when they are formally organized, and they tend to employ a much higher level of personnel responsible for distributing and monitoring activities than centrally controlled systems of the same size. Conflict resolution mechanisms are also present.

Hechter stresses the importance of homogeneous participants in minimizing errors of interpretation as to what constitutes a lawful strategy. The effectiveness of monitoring is lowered if an observed action is not clearly interpreted as either a rule-breaking or a rule-conforming act. Here again, cultural traditions are important in helping to define what is clearly within and outside the bounds of acceptable behavior. Allowing animals to trample on the sides of a canal—thereby increasing maintenance costs for everyone—may be considered either unpardonable or simply the quirk of the animals and not under the owner's control.

What constitutes an effective sanction varies from system to system. When users consider rules to be legitimate and they live in small villages where most of their future opportunities for mutual gain are based on their reputation for being trustworthy, the fear of adverse gossip alone may be sufficient to prevent most users from disregarding the rules. Many farmer-managed systems assess very small penalties on first-time offenders or those who have a reputation for respecting rules in general. On such systems, sanctions are apt to increase from an initially low level to a very high level, such as refusing water to errant farmers—or, more extremely, banishing them from the community.

In many irrigation systems run by governmental agencies, however, rule breaking may run rampant, and sanctions are imposed on those attempting to enforce project regulation rather than on those engaged in illegal behavior.[9] Harriss (1984: 322) describes the blatant rule breaking in some Sri Lankan systems where "gates are missing, structures damaged, channels tapped by encroachers and others." When asked why they did not prevent some of the more blatant offenses, two agency employees replied "that they were afraid to because of the fear of being assaulted." Risking such an assault is doubly futile considering the low probability that an offender would actually be punished.

> Prosecutions have to be carried out by the police, who have usually treated water offenses as trivial, and who do not have the same incentives to tackle them as in other cases. Further, delays over court proceedings and the very light fines which have been imposed on those who have been found guilty of irrigation offenses, have made the legal sanctions ineffectual. (Harriss, 1984: 322)

Irrigators with the appropriate connections to Sri Lankan party officials may never be prosecuted. All efforts to impose sanctions imply costs.

Devising sanctioning methods for government employees who disregard regulations is also problematic on very large projects. To sanction government employees, someone has to observe them taking illegal actions. Since the administrative staff on many of these projects is minimal in the first place, adding effective monitoring arrangements is difficult. Further, if the police and the courts already consider farmers' actions too trivial to prosecute, the illegal actions of an underpaid official accepting small bribes for special favors is unlikely to be treated very seriously. If corruption is a way of life, supervisors are likely to be unwilling to expose an employee discovered taking a bribe, unless there is a major campaign mounted against corruption and exposing an official's digression would politically benefit his or her superiors. Sanctions for simple nonperformance of work are also rare on large government projects.

Crafting Multiple Layers of Rules

The design of effective irrigation institutions affects many individuals, beginning with small groups of farmers who share a particular canal and extending outward to include many others who may not even live in the same country. Many irrigation systems are large-scale, multiple-purpose systems funded by national governments and bilateral and multilateral donors. River-basin development authorities have frequently been located on international rivers, such as the Senegal River, where the productivity of agricultural endeavors in more than a single country is simultaneously affected. The interests of diverse publics need to be considered in these multilayered systems or considerable tensions emerge as individuals seeking different outcomes attempt to interact.

The problem of crafting multiple layers of rules is exacerbated by the dominant theory of sovereignty used by policy analysts, government officials, and international donors. A theory of sovereignty assumes that a "unity of law" is necessary in all societies and that a "unity of power" is the only way to obtain this unity of law (V. Ostrom, 1988). A single center of authority is thus deemed necessary to achieve order. This center of authority is perceived as sovereign and is the maker and enforcer of all rules within a society.[10] The concept of sovereignty presumes that there can be only one source of authority in a society and that others are simply the subjects of rules determined by rulers.

> Those who have the ultimate authority to govern, and have a monopoly of the legitimate use of force in a society, exercise an authority to determine all other authority relationships. Sovereigns, then, are the source of law and cannot themselves be held account-

> able to a rule of law. All others are *subjects* in the presence of a *sovereign*; and sovereigns, not being limited to any enforceable rule of law, stand outside the law, that is, are outlaws in relation to those who are subjects. (V. Ostrom, 1988: 59; author's emphasis)

As long as national governments are perceived to be sovereign powers, economic assistance is organized on a state-to-state basis or a multilateral donor-to-state basis. Until quite recently, almost all donors worked exclusively with national governments and presumed that rules regulating irrigation would be passed in national legislatures or changed by administrative fiat in national ministries. Donor presumption of national government sovereignty, coupled with the immense flow of monetary resources from the donor community for investment in irrigation projects, has helped to increase the power of central authorities over local authorities and citizens in general.

A different concept of political order is necessary to encourage the development of multiple regimes within and across national boundaries that allow for some degree of autonomy at each level. Instead of presuming that there is one and only one source of law, it is necessary to presume that individuals at many different scales of organization can constitute their own orders as long as there are mechanisms to ensure peaceful conflict resolution. A complex multilayer polity is based on different design principles than those of a sovereign state (see V. Ostrom, 1991). Instead of authority stemming from one source alone, organization is from the bottom up as well as from the top down (see Oakerson, 1988).

Many individuals participate in crafting multiple tiers of rules, leading to a polity with extensive interorganizational arrangements in which individuals interact both horizontally and vertically. Some arrangements are "informal" in the sense that individuals undertake regular, productive activities without going through the multitudinous formal steps that are required in many developing countries to create a private or public enterprise. Beginning with the important work of Hernando de Soto (1989), far more attention has been paid to the possibility of building on an "informal" private economy to make it more vigorous in the future. Less attention has been paid to the "informal public economy," but it too may provide an initial foundation for a more vigorous and productive public sector in the future. In many societies, indigenous institutions are organized to provide public goods, but they have remained invisible to government officials and international donors.

A society, then, is not limited to only two types of institutional arrangements—the market and the state. Instead, a society can be viewed as comprising rich mixtures of private and public institutions, including local public economies (Advisory Commission on Intergovernmental Relations, 1987). In a polity composed of many interacting enterprises, the crafting of institutions is a continuous process occurring at all levels. In such a polity, conflict resolution mechanisms are more important than in polities where there is only one source of rules (V. Ostrom, 1991). If effective conflict resolution mechanisms are not present and do not recognize the relative autonomy of different levels of rule-making authority, local autonomy is apt to erode over time. Thus, in many developing countries where national governments have tended to exert their recognized power as the sovereign source of law, local rule making has occurred only in isolated locations or surreptitiously.

The diversity of attributes affecting local decision making related to irrigation makes it unlikely that any single tier of rules will be sufficient to establish mutually productive arrangements for diverse communities of individuals. From this perspective, the findings described in Chapter 1 concerning massive institutional failures in highly centralized systems are not at all surprising. We will return to this issue again in Chapter 5.

Crafting Rules in Ongoing Processes

The crafting of institutions never ends. In any complex and dynamic environment the set of rules-in-use at any point in time is unlikely to have achieved optimality. This is true even though highly motivated individuals may have crafted their own rules in the past. In a complex environment, it is difficult to ascertain which of the many factors that affect outcomes is primarily responsible for poor results. In a year when agricultural yields are poor, is it due to a shortage of rainfall, the breakdown of control gates, a new allocation rule, or increased rule breaking among participants? Similarly, if no one is willing to abide by a newly devised rule, either the rule or its monitoring or sanctioning need to be modified. Yet the causes of poor conformance to a rule are frequently difficult to discern, especially as they interact with one another. An allocation rule that would potentially help farmers to produce a better-than-expected harvest in a year with low rainfall might initially be implemented over a series of years with higher-than-average rainfall, during which its actual effect might be to lower the potential yields that could be obtained. The rule might then be rejected as unsuitable for future use, even though it might be practical for use in dry years.

The process of institutional change also involves the type of "path dependence" that characterizes technological change (David, 1988; Arthur, 1988). Historically, small changes can have a major effect on the path of innovations that is pursued because there are usually increasing returns to the use of some particular type of rule. Once one section of a large irrigation system begins to experiment with rotation, for example, the farmers in this section can begin to improve on these rules and on agricultural processes based on the expectation of their continuance. If other sections of the project adopt similar rules, they will gain even more experience and suggest more improvements. If all sections of the project adopt similar allocation rules and if the agency responsible for operating the large system is adaptable and responsive to the farmers' articulated preferences, it may be possible to adjust procedures allocating water to major canals so that they "fit" the allocation systems used in sections.[11] Over time, experience with successful rules enables individuals to learn how to use these rules even more effectively. Any effort to use alternative rules may then be doomed to rejection. Even if those alternative rules could help increase the performance of the system (once individuals gained experience with them), initial efforts to experiment with them are not likely to lead to their adoption.

Other factors also contribute to the path dependence of institutional change. As discussed above, a new rule affects not only the amount of net benefits that can be derived from a system but also the distribution of those benefits. Once some individuals have achieved a particular distribution, they will be loathe to accept a new rule that does not allocate at least as many benefits as before. This leads Freeman and Lowdermilk (1985: 101) to indicate that it is "disastrous" to make an irrigation system operational before giving serious consideration to the rules that will be used in allocating water:

> The reason is simple and profound: when water flows, some farmers are in better initial positions than others to take advantage of the resource. They quickly employ their good fortune to consolidate disproportionate advantages, and then oppose later attempts to reform the situation—usually with success because of their hold on critical resources.

Many large irrigation projects share a similar history of moving from an era of seeming abundance toward ever-greater scarcity. When a project is initiated, some farmers switch to using irrigation water, while others continue to rely on natural rainfall. The construction procedure creates a similar trajectory of behavior.

> The dam is normally built first, then the main canal is started, then the distributaries are added from the head-end downwards. Meanwhile the dam is filling while the service area is small. The top-end farmers are allowed to take and use water by methods which are very inefficient in terms of conveyance efficiency (but which save them land development and labor costs). The public authorities are more concerned that the water be used than that it be used efficiently. After several seasons the farmers' agricultural operations are "locked in" to these methods, to the point where farmers resist cut-backs in water supply which might force a higher efficiency of water use. The public authorities themselves develop patterns of behavior which reflect the priority to promote irrigation rather than rationing water. (Wade, n.d.: 7–8)

As more and more farmers begin to use water, the demand for water begins to exceed supply. The "subsequent evolution of water rights is, however, much influenced by the starting conditions in pre-scarcity conditions" (Wade, n.d.: 8). Decades of conflict may result from early developments that roughly conform to this sequence.

Because path dependence characterizes most processes of institutional evolution, all systems have limits to the degree and frequency of change that is feasible without destroying the advantages of predictable expectations created by a stable institutional process. The level of reformability that can be achieved in a set of rules varies from place to place. If the users of a set of rules do not have any *de jure* or *de facto* authority to change them, only strategic choices within a set of existing rules will be adjusted. This places a severe limit on the reformability of such systems over time. If these users have exerted at least *de facto* authority to change their own rules, the drag of the past may not outweigh all efforts to change the underlying rules. Coerced change has been attempted throughout the ages and has rarely worked very effectively. When the agents of coercion turn their backs, individuals return to their "normal" way of undertaking complex, interdependent activities.

Crafting institutions is a continuing process because of the complexity of devising institutions that match the unique combinations of variables present on any one system *and* adapt to changes in many of these variables over time. The system is never really stable. Not only are climatic conditions always variable, but the physical system tends to "wear out." In an irrigation system, dams and canals fill with silt, control structures break down, and underlying strata give way. If effective institutions are in place, considerable efforts can be devoted to counteracting physical deterioration, but no physical system operates exactly the same way year after year. As demands for water grow, conflict over water may escalate. The monitoring, sanctioning, and conflict resolution mechanisms that once were satisfactory may no longer do the job.

It is necessary to stress the ongoing nature of the process of crafting institutions, since it is so frequently described (if discussed at all) as a one-shot effort to organize farmers. Rather, those who are directly involved with the flow characteristics of a particular system, the economic conditions of a locality, and the values and norms of the users need to have *continuing* authority to craft at least some of the rules that impinge most directly on that system. Without the continuing capacity to match new rules to new circumstances, successful irrigation systems face considerable difficulties in coping with the diverse environmental and strategic threats that arise in dynamic systems.

Notes

1. These incentives are in marked contrast to those faced by irrigation officials in Korea, where parastatal organizations are responsible for irrigating 36 percent of the irrigated farmland. In each system, most of the officials were born and raised in the locality and have an economic and social background similar to that of the farmers. "So attached to the local

area are staff members that transfer out of the command area is a major threat for breach of duty" (Freeman and Lowdermilk, 1985: 106). Counteracting this attachment to a locale and resultant long time-horizon, however, is a highly centralized authority system that gives local officials and farmers little say on how irrigation systems should be operated. In the Korean case, established farmers have devised workable systems for allocating water, but they are not very efficient because control structures are poorly maintained (see Wade, 1982a).

2. Heckathorn (1984) models this as a series of nested games.

3. Since the seminal work of Walter Coward (1979), irrigation sociologists have stressed the importance of an *organizational charter* that specifies the rights and duties of irrigators and the way future decisions will be made in a legitimate and authoritative manner. A charter is a constitution for an irrigation system, specifying the rules for making collective decisions and operational choices. This is analogous to a "charter" as articulated in the U.S. Constitution (see also V. Ostrom, 1987).

4. These levels exist whether the organized human activity is public or private. See Boudreaux and Holcombe (1989) for a discussion of the constitutional rules of homeowner associations, condominiums, and some types of housing developments. See Tang (1991) for a specific application to irrigation.

5. In designing the constitution of an irrigation community, for example, setting up a legislative body requires determining how many representatives there should be. Determining the number of representatives would be affected by the physical layout. If there are five canals, having one representative from each canal may work well. If there are fifty canals, participants may want to cluster canals into branches in order to select representatives. Whatever constitutional choice is made about selecting representatives, the effect on appropriation practices results from decisions made at both a collective choice and an operational level. It is extremely hard to predict these with any certainty prior to experience in a particular setting. See the variety of rules documented in Tang (1992).

6. See Martin (1986) for detailed descriptions of the diverse allocations systems used on farmer-managed hill irrigation systems in Nepal.

7. Many of the situations where free riding could occur have the initial structure of a Prisoners' Dilemma. The task of crafting institutions is to change the incentives so that free riding is no longer the dominant strategy *or* to convert the problem into an iterated situation where one of the potential equilibria is a high level of participation and to encourage the seeking out and retention of this equilibrium (see E. Ostrom, 1990).

8. See discussion in Chambers (1980) concerning the high level of conflict that occurs within irrigation systems and the amount of time spent in conflict resolution by local leaders or administrators.

9. Government-run irrigation projects in Japan, Korea, and Taiwan are major exceptions to the lack of monitoring and sanctioning of government employees for nonperformance and illegal actions.

10. Some of the perversities of this kind of system have been elucidated by Wunsch and Olowu (1990).

11. This does not happen when the agency responsible for managing a large system has its *own* allocation system not well-matched to that used by farmers (see, for example, Reidinger, 1974).

CHAPTER FOUR

Design Principles of Long-Enduring, Self-Organized Irrigation Systems

Users and suppliers of irrigation systems must craft a variety of institutional arrangements to cope with the physical, economic, social, and cultural features of each system. Major studies throughout the world illustrate these variations in the rules-in-use (Uphoff, 1986). Even more startling is the diversity of rules used in separate branches of small self-organized systems.

Rita Hilton's 1990 study of the Karjahi Irrigation System in Nepal—a generations-old, farmer-governed system—illustrates this diversity. The small Karjahi System serves between 460 and 500 hectares and approximately 200 households. It is divided into seven *maujas* for administrative purposes, and each *mauja* makes its own rules.

> In Karjahi and Bergain, the head area always receives water first, and the tail last. In Buruwagaon, the pattern is reversed: the tail always receives water first. Gurgain *mauja* also uses a fixed pattern, but the starting point of distribution rotates annually. The plot which received water first in year "t–1" receives it last in year "t." Two additional *maujas* (Guruwagaon and Pakwai) use some sort of rotation in their areas, but the starting point of rotation is not fixed in any pattern. It is determined annually. The remaining *mauja* (Bachaha) determines the pattern of water distribution on an annual basis. The primary criterion used in setting the pattern in any one year is need: the driest plots are given water first. (Hilton, 1990: 25)

Despite the diversity of particular rules used within the specific administrative units of the Karjahi system, however, these units utilize a uniform set of design principles. This is typical of many other long-enduring, self-organized systems.

Focusing on specific rules in analyzing and prescribing institutions for irrigation systems is like focusing on specific blueprints for constructing successful irrigation projects around the world: the specific blueprints differ for each project. When local participants actively craft rules to fit their own changing circumstances over time, their rules-in-use differ also. Although blueprints vary, common engineering principles underlie the blueprints used to construct physical structures. Similarly, the rules established for particular systems are based on *design principles* that users have developed in crafting their own irrigation institutions.

Recent theoretical and empirical work on institutional design has attempted to elucidate the core design principles used in many long-enduring, self-organized irrigation institutions.[1] A design principle is an element or condition that helps to account for the success of institutions in sustaining the physical works and gaining the compliance of generations of users to the rules-in-use. A "long-enduring" irrigation system is one that has been in operation for at least several generations. Although it is impossible to evaluate the efficiency of these systems precisely, the repeated willingness of the users to invest labor and other resources is strong evidence that individual farmers receive more benefits from these systems than the costs they assume for maintaining them. It is not at all unusual for a farmer to devote twenty days of labor per year to the operation and maintenance of these systems. Farmers who divert valuable labor from other activities to dig out canal sections, repair diversion works, and operate weirs are "voting" with their backs to indicate a continued willingness to help maintain their joint facility. While all such systems impose sanctions on those who do not contribute agreed-upon resources, the size of these sanctions is sufficiently small that coercion is an unlikely explanation for system continuity. These self-organized systems thus meet the World Bank's definition of economic sustainability, even though the technical efficiency of many could be improved.

The design principles that characterize long-enduring, self-organized irrigation institutions are listed below. For these design principles to constitute a credible explanation for the sustenance of irrigation systems and related institutions, we must establish how rules characterized by such principles affect incentives.

Design Principle 1: Clearly Defined Boundaries

Both the boundaries of the service area and the individuals or households with rights to use water from an irrigation system are clearly defined.

Defining the boundaries of the irrigation system and of those authorized to use it can be considered a first step in organizing for collective action; if either of these boundaries is unclear, no one knows what is being managed or for whom. Without defining the boundaries of a system and closing it to outsiders, local irrigators face the possibility that any benefits they produce by their efforts will be reaped by others who do not contribute. Thus, for irrigators to have a minimal interest in coordinating patterns of appropriation and provision, some users have to be able to exclude other potential users from taking water.[2]

Simply closing the boundaries is usually not enough. Even those irrigators who have authorized access can abuse their privileges. Farmers at the head-end of the system may take so much water that the flow at the tail-end may be unpredictable and inadequate for agricultural use. The actual system yield may be far less than it could have been, even though some farmers have reaped considerable benefits. Consequently, in addition to closing the boundaries, rules limiting use and/or mandating provision are needed whenever water scarcity is present.

Design Principle 2: Proportional Equivalence between Benefits and Costs

Rules specifying the amount of water that an irrigator is allocated are related to local conditions and to rules requiring labor, materials, and/or money inputs.

Adding well-tailored appropriation and provision rules to boundary rules helps account for the sustenance of irrigation systems themselves. Self-organizing irrigation systems use different rules to mobilize resources for construction or maintenance and to pay water guards. In long-enduring systems, those who receive the highest proportion of the water are also required to pay the highest proportion of the costs.[3] No single set of rules defined for all irrigation systems in a region would have this result.[4] Crafting rules to apportion benefits and costs has to take into account many of the unique features of each system.

Design Principle 3: Collective-Choice Arrangements

Most individuals affected by operational rules are included in the group that can modify these rules.

Irrigation systems using this principle are better able to tailor rules to local circumstances, since the individuals who interact directly with one another and with the physical world can modify their rules over time to better fit them to the specific characteristics of their setting. Users who design institutions characterized by the first three design principles should be able to devise effective operating rules if they keep the costs of changing them relatively low.

The presence of effective operational rules, however, does not account for users following them. Nor does the fact that the users themselves designed and initially agreed to the operational rules adequately explain generations of compliance by individuals who were not originally involved in the initial agreement; this is not even an adequate explanation for the continued commitment of those who were part of the initial agreement. Agreeing to follow rules *ex ante* is an easy commitment to make. Actually following rules *ex post*, when strong temptations not to do so are present, is the significant accomplishment.

The problem of gaining compliance to rules—no matter what their origin—is frequently assumed away by theorists positing all-powerful *external* authorities who enforce agreements. In the case of many self-organizing systems, no external authority has sufficient presence to play any significant role in the day-to-day enforcement of the rules-in-use. Thus, external enforcement does not explain high levels of compliance. In long-enduring systems, however, irrigators themselves make substantial investments in monitoring and sanctioning activities. This leads us to consider the fourth and fifth design principles.

Design Principle 4: Monitoring

Monitors, who actively audit physical conditions and irrigator behavior, are accountable to the users and/or are the users themselves.

Design Principle 5: Graduated Sanctions

Users who violate operational rules are likely to receive graduated sanctions (depending on the seriousness and context of the offense) from other users, from officials accountable to these users, or both.

Now we are at the crux of the problem. In long-enduring systems, monitoring and sanctioning are undertaken not by external authorities but by the participants themselves. The initial sanctions used are also surprisingly low. Even though it is frequently presumed in modern theoretical work that participants will not spend the time and effort to monitor and sanction one another's performance, substantial evidence exists that irrigators do both in long-enduring user organizations.

To explain the investment in monitoring and sanctioning activities that occurs in these robust self-governing institutions, the term "quasi-voluntary compliance," used by Margaret Levi (1988: Ch. 3) to describe the behavior of taxpayers in systems where most taxpayers comply, is very useful. Paying taxes is *voluntary* in the sense that individuals *choose* to comply in many situations where they are not being directly coerced. On the other hand, it is "*quasi*-voluntary because the noncompliant are subject to coercion—if they are caught" (Levi, 1988: 52). Taxpayers, according to Levi, will adopt a strategy of quasi-voluntary compliance when they are confident that

> (1) rulers will keep their bargains and (2) the other constituents will keep theirs. Taxpayers are strategic actors who will cooperate only

> when they can expect others to cooperate as well. The compliance of each depends on the compliance of the others. No one prefers to be a "sucker." (Levi, 1988: 53)

Levi stresses the *contingent* nature of a commitment to comply with rules that is possible in a repeated setting. Strategic actors are willing to comply with a set of rules, Levi argues, when (1) they perceive that the collective objective is achieved, and (2) they perceive that others also comply.

Levi is not the first to stress how individuals who interact over time are able to use contingent behavior to overcome free riding (see, for example, Axelrod, 1981, 1984; Lewis and Cowens, 1983). But Levi stresses the importance of coercion as an *essential condition* for achieving the form of contingent behavior she has identified as quasi-voluntary compliance. In her explanation, enforcement increases confidence that free riding is not allowed and that those who contribute are not "suckers." As long as individuals are confident that others are cooperating and joint benefits are being provided, they willingly contribute resources to achieve a collective benefit. In Levi's analysis, enforcement is normally provided by an external ruler, even though her theory does not preclude other enforcers.[5]

Commitment in long-enduring water-user organizations cannot be explained by external enforcement. In many instances, irrigators created their own *internal* enforcement to (1) deter those who are tempted to break rules and thereby (2) assure quasi-voluntary compliers that others also comply. Given the evidence that individuals do monitor others' actions, then the relative costs and benefits must have a different configuration than posited in prior work. Either the costs of internal monitoring are lower, the benefits to an individual are higher, or both.

The costs of monitoring are low in many long-enduring systems as a result of the rules-in-use. Water rotation systems, for example, usually place the two actors most concerned with cheating in direct contact with each other. The irrigator who nears the end of a rotation turn would like to extend the time of his or her turn (and thus the amount of water obtained). The next irrigator in the rotation system waits nearby for him or her to finish and would even like to start early. The presence of the first irrigator deters the second from an early start, and the presence of the second irrigator deters the first from ending late. Neither has to invest additional resources in monitoring activities. Monitoring is a by-product of their own strong motivations to use their turn to the fullest extent. Many of the ways that work teams are organized also result in natural monitoring.

When monitoring is accomplished by an agent accountable to the other users, several mechanisms increase the rewards for doing a good job or for exposing slackards to the risk of losing their positions. In some systems, guards retain a portion of the fines.[6] All formal guard positions are accountable to the users; thus, monitors can easily be fired if they are discovered slacking off. Since users tend to continue monitoring the guards as well as one another, some redundancy is built into the monitoring and sanctioning system. A failure to deter rule breaking by one mechanism does not trigger a cascading process of rule infractions, since these other mechanisms are potentially available. Consequently, the costs and benefits of monitoring a set of rules are not independent of the particular set of rules adopted, nor are they uniform in all settings.

These five design principles enable users to constitute and reconstitute robust irrigation institutions. When users design their own operational rules (Design Principle 3) to be enforced by individuals who are local users or accountable to them (Design Principle 4) using graduated sanctions (Design Principle 5) that define who has rights and duties related to an irrigation system (Design Principle 1) and that effectively allocate the water available during different seasons of the year and other relevant local conditions (Design Principle 2), free-riding and monitoring problems are solved in an interrelated manner. Once users make contingent self-commitments to contribute, they are motivated to monitor other people's behavior, at least from time to time, in order to assure themselves that others are following the rules.

Design Principle 6: Conflict Resolution Mechanisms

Users and their officials have rapid access to low-cost local arenas to resolve conflict between users or between users and officials.

Applying rules is rarely an unambiguous task. Even such a simple rule as "each irrigator must send one individual for one day to help clean the irrigation canals before the rainy season begins" can be interpreted in various ways. Who is or is not an "individual" according to this rule? Does sending a child under ten or an adult over seventy years of age to do heavy physical work fulfill this requirement? Can someone working only four to six hours be said to have worked for one "day"? Does cleaning the canal immediately next to one's own farm qualify for meeting a community obligation? There are always ways to "interpret" the rule in order to claim compliance while actually subverting the intent. Even those who intend to follow the spirit of a rule can make errors. What happens if someone forgets about a labor day and does not appear? What happens if the only able-bodied worker is sick or unavoidably in another location?

If individuals are to follow rules over a long period of time, some mechanism for discussing and resolving what is or is not a rule infraction is necessary. If some farm families are allowed to free ride by sending less valuable workers to a required labor day, others will feel like suckers if they send their strongest workers, who could be working to produce private goods rather than communal benefits. Over time, only children and old people will be sent to do work that requires strong adults, and the system will break down. If honest individuals are unable to provide the required labor and the system does not allow them to make up for their lack of performance in an acceptable way, they will view the rules as being unfair, and conformance rates will decline.

While the presence of conflict resolution mechanisms does not guarantee that users will be able to maintain enduring institutions, it is difficult to imagine how any complex system of rules could be maintained over time without such mechanisms. In any system, land assignments and subgroup organization can increase or decrease the level of conflict facing members. When individuals hold land at both ends of a system, conflict between head and tail farmers is less severe than when no cross-cutting interests soften group antagonisms (see Coward, 1979; Downing, 1974). In many irrigation systems, conflict resolution mechanisms are informal and those who are selected as leaders are also the basic resolvers of conflict.

Design Principle 7: Minimal Recognition of Rights to Organize

The rights of users to devise their own institutions are not challenged by external governmental authorities.

This principle reflects the fact that many water-user groups organize in a *de facto* manner but are not recognized by national governments as legitimate forms of organization. Consequently, leaders of a water-user organization cannot legally open a bank account in the name of the organization or represent the interests of their members before administrative or judicial bodies. Decisions by user-group organizations may not be enforced by the police or by formal courts. Without official recognition of the right to organize, it is difficult to hold either user-group officials or members accountable for their actions.

De facto organization is sufficient in isolated locations where irrigation is used primarily for subsistence agriculture. But as soon as roads are constructed that create market opportunities for surplus products, the level of conflict over the allocation of water to different farmers or uses is likely to escalate. If government agents use their authority to support those who refuse to follow the rules of a *de facto* organization, other participants will be unlikely to continue following the rules either. An effective irrigator organization lacking formal recognition may crumble rapidly when its authority to make rules for "its own members" is challenged by the formal government.

Design Principle 8: Nested Enterprises

Appropriation, provision, monitoring, enforcement, conflict resolution, and governance activities are organized in multiple layers of nested enterprises.

Large long-enduring irrigation systems are usually organized into many tiers of nested organizations. Work teams may be as small as four or five individuals. All irrigators using a particular branch of an irrigation system may form the basis for another level of organization. A third layer may involve all farmers served by one headworks. A fourth layer may involve all systems served by the same river. If the seventh design principle holds, all these irrigation organizations would be nested in externally organized political jurisdictions (see Coward, 1979).

By nesting layers of organization within one another, irrigators can take advantage of many different scales of organization. Small-scale work teams help prevent free riding because everyone monitors everyone else. Large-scale enterprises allow systems to take advantage of economies of scale when relevant and to aggregate capital for investment. By utilizing more than a single scale of organization, many farmer-managed irrigation systems have sustained large-scale irrigation systems for long periods of time, relying primarily on their own resources—without extensive help from external agencies.[7]

Conclusion

These eight design principles are stated generally. The specific ways that suppliers and users of irrigation water have crafted rules to meet these principles vary in their particulars. Successful long-enduring institutions that appear to be based on quite different underlying designs have all developed methods to equate the costs of building and maintaining the irrigation system appropriately to the benefits that are achieved. Some examples may help the reader understand the diversity of specific rules that meet Design Principle 2.

The *Zanjeras* of the Northern Philippines. These self-organized systems obtain use-rights to previously unirrigated land from a large landowner by building a canal that irrigates the landowner's fields and that of a *zanjera*. At the time that the land is allocated, each farmer willing to abide by the rules receives a bundle of rights and duties in the form of *atars*. Each *atar* defines three parcels of land located in the head, middle, and tail sections of the service area where the holder grows crops. Responsibilities for construction and maintenance are allocated by *atars,* as are voting rights. In the rainy seasons, water is allocated freely. In a dry year, water may be allocated only to the parcels located in the head and middle portions. Thus, everyone receives water in plentiful and scarce times in rough proportion to the amount of *atars* they possess. *Atars* may be sold to others with the permission of the irrigation association, and they are inheritable (see Siy, 1982; Coward, 1979).

***Thulo Kulo* in Nepal.** When this system was first constructed in 1928, twenty-seven households contributed to a fund to construct the canal and received shares to the resulting system proportionate to the amount each invested. Since then, they have expanded the system several times by selling additional shares. Measurement and diversion weirs or gates are installed at key locations so that water is automatically allocated to each farmer according to the proportion of shares owned. Routine monitoring and maintenance assignments are allocated to work teams so that everyone participates proportionally, but emergency repairs require labor input from all shareholders regardless of the size of their share (see Martin and Yoder, 1983; Martin, 1986).

The *Huerta* of Valencia in Spain. In 1435, eighty-four irrigators served by two interrelated canals in Valencia gathered at the monastery of St. Francis to draw up and approve formal regulations to specify who had rights to water from these canals, how the water would be shared in good and bad years, and how maintenance responsibilities would be shared. The modern *huerta* of Valencia, composed of these plus six additional canals, now serves about 16,000 hectares and 15,000 farmers. The right to water inheres in the land itself and cannot be bought and sold independently of the land. Water rights are proportionate to the amount of land owned, as are obligations to contribute to the cost of monitoring and maintenance activities (see Maass and Anderson, 1986; E. Ostrom, 1990).

These three types of systems are quite different from one another. The *zanjeras* are institutional devices for landless laborers to acquire use-rights to land and water and might even be called communal systems. The *Thulo Kulo* system comes as close to allocating private and separable property rights to water as is feasible in an irrigation system. For centuries, the *huerta* of Valencia has maintained land and water rights that forbid the separation of water rights from the land being served. The Valencian system differs from both "communal" and "private property" systems because water rights are firmly attached to private ownership of land. Underlying these differences, however, is the basic design principle that the costs of constructing, operating, and maintaining these systems are roughly proportional to the benefits that the irrigators obtain.

It is important to keep these differences in mind when we discuss the application of design principles. Terms such as "privatization" may mask important underlying principles rather than provide useful guides for reform. Strict privatization of water rights is not a feasible option within the broad institutional framework of many countries. On the other hand, authorizing the suppliers and users of irrigation water to design their own systems—Design Principles 3 and 7 combined—is feasible. If participants are authorized to devise their own rules and are encouraged to learn about how others have overcome difficult design problems, we can expect motivated participants to find solutions to their own institutional problems. The proportion of successful self-organized systems can be increased if central governments invest in general institutional facilities that enhance the capabilities of those directly involved to learn new ways of governing and managing their systems, create enforceable rules, and sanction behavior contrary to these rules.

Notes

1. The methodology used to derive these design principles is contained in E. Ostrom (1990), as is the original derivation of these principles. The previous work of Coward, Chambers, V. Ostrom, Uphoff, and Wade has strongly affected my thinking on these issues.

2. Ciriacy-Wantrup and Bishop (1975) cite boundaries as the single defining characteristic of "common property" institutions, as contrasted to "open access" institutions. It is sometimes implied that this is all that is necessary to achieve successful regulation. Making this attribute one of eight, rather than a unique attribute, puts its importance in a more realistic perspective.

3. Walter Coward (1979) identified this design principle as a major characteristic of the successful irrigation systems he had examined. It was also identified by Mancur Olson (1969) as a very general principle—called fiscal equivalence—of any public institution that would achieve efficient use of resources.

4. It is sometimes argued that the rules defining common property need not be as completely specified and detailed as those defining private property. Runge (1986: 33–34) argues, for example:

> If common property—the individual right to joint use—is the norm, comparatively fewer claims must be assigned and defined. Less clarity in the assignment of rights (at least by Western standards) may also result. However, this is balanced against reduced social costs of assignment and definition.

This is true only if one means that the costs of determining the physical boundaries for individual use are eliminated and only the boundaries of the resource itself must be determined. It is not true in regard to the detailed rules that are necessary for governing how the common owners are to appropriate and provide the resource.

5. On irrigation systems that are owned and operated by government agencies, the agency could also provide the type of monitoring and sanctioning Levi has in mind. Robert Wade (1987) has a similar view of the willingness of many irrigators to comply with reasonable rules if they were assured that others would also comply and that those who did not would be sanctioned.

> In many situations individual irrigators will restrain their water rule breaking *if* they are confident that others will also refrain and *if* they are confident that they will still get as much water as they are fairly entitled to (even if not as much as they would like). They will more likely refrain from cheating if they are confident that by doing so they will not be the "suckers." Where people are motivated by an "I'll restrain if you restrain" calculation, then an institution (such as an irrigation department) that convinces them that these expectations are justified can promote voluntary compliance with the rules. (Wade, 1987: 178; author's emphasis)

6. In some systems, guards are paid a proportion of the crop at the end of the year. With this type of payment, the guard's own payment depends on keeping the reliability of the system as high as possible so that farmers being served can produce as much as possible.

7. See Maass and Anderson (1986), Siy (1982), and Pradhan (1989a, 1989b) for descriptions of larger and more complex irrigation systems relying on nested organizational arrangements.

CHAPTER FIVE

Applying Design Principles

The design principles discussed in Chapter 4 were derived from analyses of self-organized, long-enduring irrigation systems located in many countries. Many of these systems now operate within sophisticated, multilayered institutions crafted over long periods of time, even though their physical structures are relatively simple. Long-term survival should not be equated with optimal performance, even though it demonstrates sustainability. Self-organization does not guarantee that optimal institutions will be crafted.

Matching rules to local circumstances is often difficult. Not all systems find a set of rules that adequately meets the problems they face. These systems either limp along with constant conflict and insufficient resource mobilization, or they do not survive at all. Previous investments in physical and social capital are wasted, and farmers return to dry-land agriculture, producing yields far below what they could with irrigation.

Because institutions are invisible, it is not obvious to external observers whether a particular farmer-organized irrigation system has crafted rules that meet the design criteria described in Chapter 4. What they can see are the temporary diversion works, unlined canals, and the lack of modern control mechanisms that characterize so many farmer-organized irrigation systems. Both successful and marginal farmer-organized systems appear primitive and ineffective to an engineer who expects to see permanent diversion weirs, lined canals, and effective placement of all physical works.

External technical assistance and better physical works can improve the efficiency and agricultural yield of farmer-organized irrigation systems. Significant increases in yields can also be achieved by improving the operation of extant systems (see Chapter 1). Attempts by external agencies to assist farmer-managed irrigation systems, however, have also adversely affected performance.

Analyses of these failed attempts have pointed to a lack of awareness by project designers of the institutions that already existed (Coward, 1985). Project designers of unsuccessful reconstructions frequently assumed that nothing of any value existed before the physical works that they had planned. The amazing number of successful reconstructions of farmer-organized irrigation systems in the Philippines and in Nepal attests to the potential for improvement of these systems when project designers are aware of existing institutions and farmers are directly involved in the design of new physical works and the institutions for financing and operating these systems (see F. Korten and Siy, 1988; Pradhan, 1989b).

The need to apply institutional design principles is even more pressing when we examine those large government-owned irrigation systems that have proved unsustainable. Many of these systems have permanent diversion works, lined canals, and modern control mechanisms. But as discussed above, little maintenance has been undertaken, and the level of conflict, fear, and suspicion among farmers is substantial. Crafting improved institutions on these new systems is significantly more difficult than improving the operation of existing farmer-organized systems.

In most of these large-scale systems, few of the design principles discussed in Chapter 4 are met even to a minor degree. Service area boundaries are somewhat vague in practice, and no one is quite sure who obtains water. The farmers being served pay only a small proportion, if any, of the costs involved in construction, operation, and maintenance. Neither the farmers nor the government officials involved in the day-to-day operation participate in crafting system rules. No one's behavior is monitored or sanctioned, and few conflict resolution mechanisms are available. Where farmers *are* formally encouraged to organize, officials insist that everyone follow the same organizational blueprint.

The design principles are potentially powerful tools for *diagnosing* and *explaining* why some irrigation projects are not sustainable. They can also be used for *prescribing* reforms as long as such proposals presuppose that reform is an ongoing process that must involve water users. Reforms based on these design principles may, however, generate considerable opposition. For example, Design Principle 2 (along with the general approach outlined in this study) requires beneficiaries of irrigation projects to cover at least the recurrent costs of those projects. Proposals consistent with this design principle have frequently met strong resistance. If such opposition is not anticipated and understood, reform proposals applying these design principles have little chance for long-term implementation.

For this reason, the next section analyzes recent financial support for irrigation projects and the sources of resistance to changes in these financial institutions. The following section reviews the experience of one long-term effort to achieve reform using the eight design principles. The final section recommends specific strategies for donor agencies and host governments to enhance the performance of irrigation institutions.

Financial Incentives and Irrigation Institutions

A frequent source of opposition to reform stems from the way large irrigation projects—and even some small farmer-managed projects—have been funded.[1] Funds for constructing, operating, and maintaining systems typically come from the taxpayers of the nation in which the irrigation system is located or from the taxpayers in those nations providing economic assistance. Hence, the financial connection between supply and use is lacking. Whether the resources so mobilized are directly invested in the construction and operation of irrigation systems or are diverted for individual use by politicians or contractors depends on the professionalism of those involved and on active efforts to monitor and sanction the diversion of resources. When the eventual users are involved in construction and operation, they provide low-cost monitoring of how resources are used. When the users are not involved, expensive auditing systems are needed but are rarely supplied. Consequently, a considerable portion of the mobilized funds is diverted to purposes other than those for which it was intended.

Further, project design is aimed more toward capturing the approval of those who fund new construction than toward providing systems that solve the problems facing present and future users. To convince politicians that large portions of a national budget should be devoted to the construction of irrigation projects, planners attempt to design projects that are "politically attractive." This means that politicians who support such expenditures can claim that the voters' funds are being invested in projects that will greatly expand the amount of food available and thus lower the cost of living.

To convince external funding agencies that major irrigation projects should be funded through loans or grants, the evaluative criteria used by these agencies in selecting projects has to play a prominent role in project design. Engineers, who generally lack experience as farmers or training as institutional analysts, often aim toward winning political support or international funding. As a result, their projects may fail to serve most small farmers effectively and thus discourage users from investing in the long-term maintenance of projects. Inefficiencies occur at almost every stage. At the same time, this inefficient process leads to the construction of projects that generate substantial profits for large landholders and strong political support for a government.

All types of opportunistic behavior are encouraged, rather than discouraged, by (1) the availability of massive funding to subsidize the construction and operation of large-scale irrigation projects and (2) the willingness (or even eagerness) of national leaders to subsidize water as a major input for agricultural production. Corrupt exchanges between officials and private contractors are a notorious and widespread form of opportunism; corrupt payments by farmers to irrigation officials are less well known but probably no less widespread. Free riding on the part of those receiving benefits and the lack of trust between farmers and officials, as well as among farmers, are also endemic. Further, the potential rents that can be derived from free irrigation water by large landowners stimulate efforts to influence public decision making as to where projects should be located and how they should be financed. Politicians, for their part, win political support by deciding who will benefit from artificially created economic rents.

Robert Bates explains many of the characteristics of African agricultural policies by arguing that major "inefficiencies persist *because* they are politically useful; economic inefficiencies afford governments means of retaining political power" (Bates, 1987: 128). Part of Bates's argument relates to the artificial control exercised over the prices paid for agricultural products, a topic not addressed in this study. The other part of Bates's argument relates to the artificial lowering of input prices.

> When they lower the price of inputs, private sources furnish lesser quantities, users demand greater quantities, and the result is excess demand. One consequence is that the inputs acquire new value; the administratively created shortage creates an economic premium for those who acquire them. Another is that, at the mandated price, the market cannot allocate the inputs; they are in short supply. Rather than being allocated through a pricing system, they must be rationed. Those in charge of the regulated market thereby acquire the capacity to exercise discretion and to confer the resources upon those whose favor they desire. . . .
>
> Public programs which distribute farm credit, tractor-hire services, seeds, and fertilizers, and which bestow access to government-managed irrigation schemes and public land, thus become instruments of political organization in the countryside of Africa. (Bates, 1987: 130)

So there is an added dimension to rent seeking in many developing countries: the losses that the general consumer and taxpayer accrue from rent-seeking activities and the acquisition of resources needed to accumulate and retain political power. All forms of opportunistic behavior are therefore exacerbated in an environment in which an abundance of funds is available for the construction of new and frequently large-scale irrigation projects that provide subsidized water. This is exactly the political and financial milieu that irrigation suppliers have faced during the past forty years in most developing countries. Developed countries have made vast amounts of money available to developing countries through bilateral and multilateral loans and aid agreements.

By comparison with the large sums of money that have been available for the construction of irrigation projects, official fees collected from farmers served by government-operated irrigation systems in many countries have been minuscule. A recent study of the official revenue received from farmers in Indonesia, Korea, Nepal, the Philippines, Thailand, and Bangladesh indicates that only in the Philippines do the fees collected equal or exceed the costs of operating and maintaining the systems. But none of these countries collected enough to meet a small proportion of amortized capital costs (Repetto, 1986: 5). The actual "price" that farmers may pay in illegal bribes is far from minuscule on some projects; however, these "fees" are not reflected in public records, nor are they used for the operation and maintenance of irrigation systems (other than as bribes to low-level employees that are far larger than their official paychecks). The amount of bribes and the fees paid to private tube-well operators demonstrates the farmers' willingness to pay far more than the current subsidized price for reliably available water. Farmers also derive higher agricultural yields when served by private rather than public suppliers because the water supply is more reliable (Repetto, 1986: 7).

Many analysts view the financial largesse for designing and constructing new irrigation systems, combined with the lack of funding for operation and maintenance, as the major cause of the severe problems facing irrigation projects in developing countries. Changing the rules linking the supply of funds to the use of water is a frequently cited priority for reform (Easter, 1985; Repetto, 1986; Small et al., 1986; Wade, 1987; O'Mara, 1989), but it is not uniformly supported by researchers who have spent long periods in the field observing irrigation systems (see, in particular, Moore, 1989). Donor agencies have urged national governments to commit themselves to a major change in the way that irrigation is financed, but donor agency staff also face incentives that deter them from taking a strong stand to recapture recurrent costs, let alone capital costs. Much of the focus on their performance ratings concerns the facility with which they move large quantities of money and manage projects. The well-known winning strategy for meeting these performance criteria is to approve a small number of very large capital-intensive projects (see discussion in Tendler, 1975; E. Ostrom, Schroeder, and Wynne, 1990). In addition, donor agency personnel are often assigned to a particular country or region for a relatively short term. Although many donor-supported projects are funded with the contingency that beneficiaries pay user fees to finance recurrent costs, the short tenure of donor agency personnel precludes the tenacity needed to ensure that this contingency is actually met. New personnel who are unaware of this commitment are transferred into the locality; meanwhile, the system has fallen into disrepair for lack of funding. The obvious need for reconstruction leads new officials to approve yet another reconstruction—a large, capital-intensive project. The ease with which it has been possible to obtain funds for reconstruction of major projects that were not maintained has sent confusing signals to host governments as to how serious donors really are about the need to reform the financing of irrigation systems (or other major infrastructures).

Proposals to increase user fees on government-owned irrigation systems, however, meet virulent opposition from farmers, politicians, and irrigation officials. International donors have long argued that national irrigation agencies should charge fees that at least cover recurrent costs, if not some of the capital costs as well. It is easy to understand why farmers would oppose increases in the official fees they are supposed to pay. The economic rents obtained from artificially low input costs are rapidly capitalized back into the value of the land when it has access to cheap water. Hence, a change in fee structure means that not only will farmers have to pay substantially more for water but land values will fall as well. Landowners with access to subsidized water are able to capture much of the artificial rents in the price they charge a tenant. But the tenant is likely to be the person who has to pay the increases in water fees.

Farmers' resistance to increased fees has an objective basis. If the fees charged for water on some projects were to be raised sufficiently to cover *both recurrent and capital costs*—and the higher fees were actually enforced—many farmers would be better off not irrigating. They could not earn enough money from enhanced yields to cover the marginal costs of the higher irrigation charges. A recent study examining the feasibility of imposing water charges to cover full costs in Indonesia, Korea, Nepal, the Philippines, and Thailand, concludes that "the benefits of irrigation are not great enough to make possible the full recovery of costs in any of the five countries without making farmers worse off than they were before the introduction of irrigation" (Small et al., 1986, cited in Repetto, 1986: 8). In other words, the total benefits generated by these projects are not, in practice, greater than the costs of the projects. Farmers understandably resist paying for the excesses of the past.[2]

The situation is only slightly better when one contemplates fees that cover *recurrent costs alone* without attempting to recover all past capital investments. On one hand, the same study concluded that the aggregate benefits derived from irrigation projects in the five countries listed above are sufficient that farmers could afford to cover recurrent costs. But, even here, farmers have objective concerns. Aggregate benefits and costs average out the highly variable performance of different projects. In actuality, the benefits obtained from irrigation on some projects may not entirely cover even the recurrent costs. Further, water fees are not tied to system performance. If fees are not related to the availability and predictability of water, farmers may be asked to pay for water they never receive. In many developing countries, water fees are used as general income by the national governments and are not actually allocated to irrigation agencies. Irrigation agencies, therefore, do not depend on the collection of fees for their operational income. But when irrigation agency personnel fail to respond to farmers' concerns unless their palms are greased with bribes, farmers are understandably hesitant to pay for water over which they have no control.

Whether farmers on a particular project are sufficiently better off as a result of increased agricultural yields is highly problematic. Actual returns to the farmer depend on the price received for the agricultural yield; the price and availability of necessary inputs including credit, new-variety seeds, and fertilizer; and the fees charged for water. A 1980 economic analysis of the potential return to farmers from the BICOL Integrated Area Development, for example, concluded that some farmers would be substantially worse off if proposed fee increases were imposed. In particular, those farmers who had previously irrigated their lands using small-scale gravity-fed systems would be worse off under the new fee-supported system unless prices for their product radically increased or farm yields were to exceed those already achieved by the more productive farmers in this region.[3]

All of the following must occur before farmers are sufficiently better off so that payment of fees covering the recurrent costs of many projects is objectively feasible:

1. Farmers must have confidence that water will be available whenever it is needed before they will either (a) invest in expensive inputs related to a single crop and/or (b) make such investments in regard to double or triple cropping.[4]

2. Farmers must be able to obtain credit at a reasonable interest rate in order to purchase more expensive inputs.

3. When new inputs are needed, farmers must be able to obtain them at market-clearing prices.

4. Farm income must exceed the increased costs of new inputs.

5. The increased net returns must be greater than the O&M costs assessed against the farmers.

Unless the first four conditions are met, farmers served by irrigation infrastructures will not invest in the inputs that are necessary to generate increased agricultural yields. Unless the fifth condition is met, farmers will strongly resist paying monetary fees or volunteering their labor for maintenance activities.

Farmers' resistance to paying fees that cover recurrent costs has many long-term consequences for the sustenance of major irrigation projects. Unless farmers pay the fees used to hire O&M staff or they perform these O&M activities themselves, many irrigation agencies will not be able to do anything more than operate systems in a minimal fashion. Little investment can be made in routine or emergency maintenance. The initial lack of maintenance triggers a vicious circle that has been characteristic of many large systems constructed in recent years. Without adequate maintenance, system reliability begins to deteriorate. As reliability diminishes, farmers are less willing to make investments in expensive seeds and fertilizers that are of little benefit without a reliable water supply. Without these input investments, the net return from irrigated agriculture declines. As returns fall, farmers become still more resistant to contributing to the system's sustainability.

The Philippine Experience with an Ongoing Process of Institutional Reform

Breaking out of these vicious circles is extremely difficult. The process undertaken by the National Irrigation Administration (NIA) in the Philippines is one example of a reportedly successful effort to develop different rules for financing recurrent and capital costs. The Philippine experience is noteworthy for many reasons. First, the participants were conscious of the need to adopt a learning approach rather than a blueprint approach (see D. Korten, 1980). Second, many rules affecting finance, design, construction, maintenance, and use were changed (we will only discuss the changes related to finance). Third, these rule changes led to well-documented improvements in system performance. Fourth, considerable effort was devoted to increasing other aspects of social capital, including the skills and understanding of irrigators and public officials. Fifth, opposition to these reforms from within NIA, resulting from the potential loss of jobs and power, stopped the momentum of change at several junctures.

In the early 1960s, the Philippine government contemplated a major irrigation program directed toward achieving self-sufficiency in rice production. It created the NIA as a semiautonomous corporation with broad powers to undertake irrigation development. Initially NIA received a large subsidy from the Philippine national government to cover both construction and O&M. The understanding, however, was that NIA would eventually become self-financing. The first step was that NIA should cover its own recurrent costs. Yet, as Benjamin U. Bagadion (a key participant in the evolution of a new set of irrigation institutions) explains, NIA was far from being able to cover its own recurrent costs, let alone construction costs. During fiscal year 1964–65, "irrigation fee collections totaled only 1.27 million pesos [$0.33 million in 1964 U.S. currency] while operation and maintenance expenses were 3.42 million pesos [$0.88 million in 1964 U.S. currency]" (Bagadion, 1988: 7; currency conversions added). In 1967, NIA attempted to solve the O&M budget deficit on national systems by increasing irrigation fees substantially. The results were counterproductive.

Although total collections increased, expenses also rose as efforts were made to upgrade operation and maintenance to justify the higher fees. Consequently, NIA's net budget deficit remained. Moreover, the percentage of collectible fees actually paid decreased from 59 percent before the rate increase to 27 percent afterward. With no solution in sight, the government continued to provide the subsidy, and NIA's O&M problems did not receive meaningful attention for another half decade.

A similar failure occurred in an early effort to create irrigators' associations to manage smaller systems that NIA wanted to return to farmer control. "Paper" associations were created, but did little other than fulfill legal requirements. Farmers were not consulted about proposed changes on their systems, and they saw no reason to assume responsibility thereafter. In addition, "farmers knew they could lobby their member of Congress for additional free 'pork-barrel' assistance, so they often let their system fall into disrepair, waiting for the government to do the work" (Bagadion, 1988: 7).

In 1974, NIA's charter was substantially amended to enable it to operate more like the public corporation it was intended to be. Before this time, fees collected by NIA were remitted back to the national treasury. The regular budget of the agency was included as part of the general appropriations procedures. The amended charter allowed NIA to keep the irrigation fees it collected, while providing for a subsidy to explicitly cover O&M and new construction costs for both national and communal systems.

> The new arrangements created a potential incentive for NIA personnel to focus on collections—the more funds collected, the more the NIA would have available for the operation and maintenance of its systems. Paradoxically, the very amendment which provided for an explicit subsidy also allowed the NIA to begin to gear itself for the eventual removal of the subsidy. The understanding with the government budgetary authorities was that the subsidy for operation and maintenance expenses was to be gradually phased out over a period of five years. The NIA would then be directly dependent upon collections from farmers for all of its operation and maintenance expenses. (Bagadion, 1988:8)

For national irrigation systems, the previously unenforced policy of requiring a payback of construction costs over twenty-five years was changed to a policy of recapturing over a fifty-year period without interest. For communal irrigation systems, "the new policies neutralized the adverse effects of the 'pork-barrel' system in which communal irrigation facilities were built without any recovery of costs from the farmers, a system which had fostered the associations' dependence upon the government" (Bagadion, 1988: 9). David Korten (1988: 137) indicates that this change meant that farmers were "no longer welfare clients accepting whatever their benefactor chose to offer, but rather were customers buying a service with the option of withholding agreement and/or payment."

Although the foundations for giving legal status to irrigation associations were already in place, actually organizing these associations after years of strong central control over irrigation was not easy to accomplish. It took the creative energies of many inspired public officials, newly hired irrigation organizers, and devoted academics and solid support from the Ford Foundation to organize strong user associations that could relate effectively with an all-powerful supplier like the NIA.[5] Simply changing the financing rules of NIA—the supply side—without strengthening the authority and skills of the users was not sufficient, nor would efforts to improve the use side have worked without changes in the supply side. Changes on both sides are usually critical to the success of any institutional reform.

A key change related to the budgeting and appropriations procedures was adopted. Under the old system, the budgetary year began on January 1, but, as in many other countries, funds often were not released until three months into the budgetary year. No construction could be undertaken during the first three months, yet these are the dry months, which are ideally suited for construction (D. Korten, 1988: 129). In 1979, a new budgetary rule made things even worse by requiring unexpended funds to revert to the national treasury. Construction of irrigation projects frequently came to a screeching halt at the end of December, remained idle during the dry months, was damaged by typhoon rains, and had to be rebuilt before the projects could be completed during the next year. Construction costs were higher than necessary, and commitments to farmers could not be kept with much assurance. This problem was eventually solved by a series of steps to change the way in which funds were appropriated and expended.[6]

During the early 1980s, NIA's subsidy for recurrent costs was slowly withdrawn. Each provincial office was urged to determine the amount of new communal construction that would be necessary to obtain sufficient revenue to pay for the provincial operations budget. "The average province required an area of 3,000 to 4,000 hectares with *satisfied* irrigators making regular amortization payments" (D. Korten, 1988: 137; emphasis added). While all income was deposited in one general account, records of costs and revenues were kept by province, allowing officials to keep track of net flows.[7] These changes in financial rules made NIA staff focus on fiscal solvency as the bottom line. But at the same time, Korten reported, they learned "that the way to achieve financial viability was to stay close to the customers and provide satisfactory service."[8]

Several lessons can be derived from the Philippine experience. First, simply raising irrigation fees without finding better methods of relating supply to use did not work. It was, in fact, counterproductive. Second, it took many changes in rules, some of which were relatively small and subtle in nature, to have a major impact on the actual incentives facing agency staff. Third, many of the rule changes affected supplier incentives related to system design, construction, operation, and maintenance. Fourth, improvements in performance came slowly. Fifth, agency personnel resisted internal changes. Sixth, in addition to the work of devoted public servants, external help in the form of intellectual capital and financial support played an important role in the crafting process. Seventh, the process of change focused more on communal and small national systems than on the large national system.[9] And eighth, the process of crafting effective institutions never ends.

This brief overview of the Philippine experience helps us understand why farmers, politicians, and irrigation staff oppose substantial changes in the budgetary practices that predominate in many countries. Farmers can be counted on to vigorously oppose proposals to raise fees because increased fees rarely carry believable promises to enhance system performance. Politicians lose one source of power when irrigation is no longer a part of the "pork-barrel" politics of a nation; hence, politicians are unlikely to initiate major changes in fee structures unless pushed hard by tight budgetary constraints. It is far more difficult for irrigation engineers to spend time and energy meeting with farmers and worrying about the financial solvency of their agency than it is for them to receive a guaranteed income no matter what they do. Finally, if changes in financing eventually result in the transfer of system operation and maintenance to the irrigators, O&M personnel will lose their jobs. Bagadion (1988: 18) reflects that the displacement of NIA field-level personnel was an important problem in the Philippines that "slowed the expansion of the participatory program in national systems." Thus, proposals for major changes in fee structure are likely to come only as a result of extreme budgetary restraints reinforced by external donor insistence.

Recommendations for Enhancing the Performance of Irrigation Institutions

Citizens, government officials, external donor agencies, and others seeking to improve irrigation institutions can gain valuable insight from this experience in the Philippines: any attempt to achieve meaningful improvements in complex institutional arrangements that currently generate considerable benefits for powerful and well-organized individuals will take a long time and considerable work. Boss Plunkitt of Tammany Hall was famous for his insight that "reformers were only Morning Glories" (Riordon, 1963). Those who try to reform systems that generate substantial rents for powerful and well-organized interests must recognize that those rents will be used to avoid reform. It takes considerable will, work, and perseverance to avoid blooming early in the process but wilting when the opposition gets tough. Simple pronouncements by donors or central governments will not accomplish major reforms.

Reforms involving user fees, such as those frequently proposed in the literature, will always generate extreme opposition. At the same time, several types of institutional reform based on the design principles presented in this study are both essential and somewhat less likely to be the source of strong opposition. The first strategy relates to the establishment of authority for user groups of various types to create their own corporate entities. This authority was already in place in the Philippines and was one of the building blocks used in that experimental program. This authority is similar to that of a group of individuals to establish a private corporation to achieve legal objectives. Private corporations can create their own charters in some countries as long as they meet certain overall specifications. If those who wish to organize to achieve a public purpose can rely on general authorization to create their charters, the seventh design principle can be achieved at lower cost. To be a recognized user group, a group might need to open its books to all members, allow some form of external auditing, and recognize the rights of all citizens to information about system performance. Examples of successful user-group charters might be used in training programs to illustrate the types of rules used in the more successful systems.

The second strategy for institutional reform relates to investments in courts and other forms of conflict-resolution mechanisms. Without a fair, low-cost, general purpose court system, it is extremely difficult to craft institutions that solve difficult problems. While those directly involved may be willing to take on substantial responsibility for monitoring and sanctioning activities, some conflicts are likely to escalate and need resolution by external, impartial, and fair officials.

Considerable opportunity for reform exists in efforts to improve the performance of small farmer-owned irrigation systems. Many of these already have effective farmer organizations. Many do need better physical capital and knowledge about how to improve agricultural yields. *Institutional Incentives and Rural Infrastructure Sustainability* (E. Ostrom, Schroeder, and Wynne, 1990) makes some specific recommendations concerning strategies that could be adopted for small irrigation projects. This advice bears repeating here.

One likely point for intervention in small-scale projects is when external assistance is requested. Donors and national governments who are interested in enhancing investments in sustainable, small-scale projects should assist these groups *only* when firm evidence exists that those who are supposed to benefit from a facility

1. are aware of the potential benefits they will receive

2. recognize that these benefits will not fully materialize unless facilities are maintained

3. have made a *firm commitment* to maintain the facility over time

4. have the organizational and financial capabilities to keep this commitment

5. do not expect to receive resources for rehabilitating the facility if they fail to maintain it

This can be accomplished by investing in infrastructure projects that meet the following conditions:

1. The direct beneficiaries are willing to invest some of their own resources up front.

2. The direct beneficiaries are willing to pay back a substantial portion of the capital costs (at low interest and over a long time, if necessary) and to undertake maintenance.

3. The direct beneficiaries are assured that they can
   - participate in designing the project
   - monitor the quality of the work performed
   - examine the accounts that form the basis for their financial responsibilities
   - protect established water rights
   - hold contractors accountable for inferior workmanship that is discovered after the system is in operation
4. The granting agency is assured that
   - farmers' commitments to repay costs will be enforced by appropriate legal action, if necessary
   - farmers have an effective organization with demonstrated capabilities to mobilize resources, allocate benefits and duties, and resolve local conflicts
5. All donors and the host government are firmly committed to the above principles and will not provide funds to bail out those beneficiaries who fail to perform their responsibilities.[10]

Individuals who are willing to make initial investments to obtain capital goods demonstrate their own recognition of future benefits. Furthermore, the higher the proportion of the capital investment that beneficiaries are willing to repay, the greater the likelihood that the beneficiaries will attempt to make economically feasible investments to enhance productivity rather than seek rents. If the infrastructure is really going to increase the well-being of the supposed beneficiaries, they will have increased resources to devote to repayment. Furthermore, if beneficiaries know that they have to repay capital costs, they are likely to insist (if they have the institutional autonomy to do so) that the project have a high probability of producing net benefits. Under these conditions, donor or central government funds support projects that are considered to be of real value to the participants.

This means that direct beneficiaries or their representatives must be involved in the design and financial planning of an infrastructure producing highly localized benefits, and they must have the right to say no to a project that they do not think is worthwhile. If they cannot say no, they cannot make a commitment that is considered binding because they can always assert that they were forced to agree. In addition, to make enforceable commitments, the beneficiaries need to be

- organized in a legally recognized form prior to the creation of financial and construction arrangements. Beneficiaries can then participate in the design and financing of the project, as well as in the approval of a contract to eventually assume ownership of the facility and responsibility for its maintenance.

- confident that government officials are also making enforceable contracts—that beneficiaries can hold public officials accountable as well as being held accountable themselves.

- assured that future conflicts over contract enforcement will be resolved fairly and that impartial conflict resolution arenas exist if needed. (E. Ostrom, Schroeder, and Wynne, 1990: 152–53)

Efforts to craft new institutions to improve the performance of recently constructed large-scale, government-owned irrigation projects will be more difficult to accomplish than efforts to improve small-scale projects (see Tang, 1992). Farmers have to learn how to trust other farmers and irrigation officials. Substantial changes are usually needed in the overall management of the system. Irrigation officials are not likely to be very responsive to farmers' requests to meet schedules when the farmers refuse to pay irrigation fees. All the problems that occur on large systems cannot be solved simultaneously in a short time. Consequently, officials should hire well-trained field workers who can work directly with farmers and system engineers.[11] Reform efforts will require decade-long perspectives rather than the more typical time-horizon of a budget year or the current crop. Institutional reform is a long-term investment in social capital.

Notes

1. See Niasse (1990) for an analysis of the perverse incentives involved in the initial design of "irrigated village perimeters," or PIVs, in the Senegal Valley during the 1970s. As long as government subsidies were available for essential inputs and drought conditions continued predisposing farmers to irrigation, PIVs multiplied at a substantial rate. Since the Senegal government accepted the structural adjustment programs of donor agencies, more actual costs are being borne by the farmers. Given the substantial capital costs involved, more and more land is now left idle and agricultural productivity in the region is dropping precipitously.

2. Thus, a key problem facing policy makers in many countries is how to make the best economic use of projects that were poorly designed in the past. If it is impossible to recover costs from the profits made by farmers on agricultural products, there is no economic justification for continuing to operate a project. Many projects that do not currently recover full costs could be governed and managed so as to do so in the future (plus some contribution toward recovery of capital investment).

3. The average income of a farmer able to irrigate his land from the previously existing small-scale systems was 3,819 to 3,943 pesos ($519.38 to $536.25 in 1979 U.S. currency) per year for an average farm of 1.65 hectares. With an irrigation fee of 18 *cavans* of palay rice (then being proposed), the average income for such a farm would drop to 2,747 to 2,871 pesos ($373.6 to $390.46 in 1979 U.S. currency) per year. With a fee of 12 *cavans* of palay rice, the average income would be 3,242 to 3,366 pesos ($440.91 to $457.77 in 1979 U.S. currency) per year. Alternatively, if the farmer were able to increase his yields above that which had already been achieved in the area or were to receive a higher price for rice, economic returns could be higher even with the proposed irrigation fees (see Sommer et al., 1982: Appendix D).

4. Reliability of the water supply can be achieved by a combination of physical and institutional means, but it is difficult. Unless sufficient storage is available in the system, the demand for water is limited, and effective physical regulation of the system is built into the designs, the potential for extremely high levels of conflict among farmers and between farmers and irrigation agency officials is always present. If a set of institutional rules for allocating water is understood, accepted as legitimate, implemented, and enforced, conflict over the allocation of water can be reduced and reliability achieved. This need for effective allocation rules has been ignored in the design of many major irrigation systems in recent times.

5. Furthermore, the individuals involved have written extensively about their experience, providing a means for others to gain general knowledge from the particular experience. David Korten had already developed a strong theoretical argument for learning by doing and keeping good process documentation of various experiments so that experiments did form the foundation for cumulative understanding (D. Korten, 1980). The recent book edited by Frances Korten and Robert Siy (1988) synthesizes the reflections of key actors in this learning process.

6. First, they obtained "a change in the appropriations process so that the appropriation for communals was made on a lump sum basis rather than on an individual project basis" (D. Korten, 1988: 130). This gave them more flexibility to shift funds among projects and a greater capacity to keep commitments made to user groups. Then, Korten writes, the NIA began to draw on its corporation fund: "By 1980 this fund had become substantial and the NIA began to use it to finance communal construction work during the initial three months of the year, pending release of the new annual appropriation." Repayments were made once the appropriations had been released. "The problem of returning unexpended funds to the national treasury at the end of the year was eventually solved by appropriating the communal irrigation funds to the Ministry of Public Works instead of directly to the NIA." When the Ministry released funds to NIA, they were legally "expended" and did not have to be returned, Korten concludes.

7. Provinces that had an excess of revenues over their expenditures received an incentive payment of 10 percent of their surplus with considerable discretion as to how to spend these funds, including limited incentive bonuses to staff. Financial performance at the provincial level was built into the staff performance ratings. Irrigation staff learned that it was "difficult to collect from farmers on projects that had been unsuccessful in increasing production, where the facilities constructed were inoperable or where antagonistic relations with the farmers had developed" (D. Korten, 1988: 137–38).

8. The Indian state of Maharashtra has been able to achieve a relatively good record for collecting irrigation fees from farmers as well. A recent study summarized in Easter (1985: 22) found that "fees collected were 66, 62 and 89 percent of the O&M costs in the minor, medium and major irrigation system respectively." Easter lists four major factors in successfully collecting water fees:

- government sanctions on farmers not paying water charges, when they apply for irrigation water each year
- fines for nonpayment of water charges by a fixed date

- good irrigation service

- good communication among irrigation officials and farmers

9. As Bagadion (1988: 18) notes:

> While touching all of the provincial and regional offices of the NIA, including all of NIA's communals work and some of its work on the small and medium sized nationals, improvements in these programs are still needed, and change has yet to come to the larger national projects and systems. The processes used in small and medium national systems need to be applied more widely and creative thinking is needed regarding the application of such processes to larger systems.

10. In light of the imperative that donor agency officers "move money" and the temptations of rent seeking for government officials, this is a particularly difficult commitment for donors and host governments to make. It may require the major donors to work together with the host government on a joint funding strategy. Both donors and host governments may want to provide funds in case of major disasters to help rebuild structures destroyed by earthquakes, floods, and avalanches. This is a form of "insurance" that does not destroy incentives to undertake routine maintenance unless the definition of an externally caused disaster is interpreted too broadly.

11. Uphoff (1986) provides an excellent summary of the problems involved in successfully changing the patterns of interactions on large-scale irrigation systems. The efforts of an Agrarian Research and Training Institute (ARTI)/Cornell team on the Gal Oya project in Sri Lanka are illustrative of the type of intervention that is likely to be needed. Field-workers who were college graduates but came from farming families were employed as organization "catalysts" that could help farmers begin to solve some of the more immediate and small-scale problems without any need for a formal organization. By building confidence that joint problems could be solved, these field-workers helped farmers build trust in one another. By communicating farmers' needs to irrigation officials and helping to change the way the larger system operated, further trust was built. Such approaches require substantial investments in personnel who are willing to undertake this perplexing and difficult work. The potential benefits that can be achieved, however, are substantial.

References

Advisory Commission on Intergovernmental Relations 1987. *The Organization of Local Public Economies*. Washington, D.C.: Advisory Commission on Intergovernmental Relations.

Alchian, Armen A., and Harold Demsetz. 1972. "Production, Information and Economic Organization." *American Economic Review* 62 (5): 777–95.

Ali, Syed Hashim. 1985. "Planning and Implementation of Measures to Ensure Productivity and Equity under Irrigation Systems." In *Productivity and Equity in Irrigation Systems,* ed. Niranjan Pant. Dehra Dun: Natraj.

Arthur, W. Brian. 1988. "Self-Reinforcing Mechanisms in Economics." In *The Economy as an Evolving Complex System,* ed. Philip W. Anderson, Kenneth J. Arrow, and David Pines, 9–31. Reading, Mass.: Addison-Wesley.

Ascher, William, and Robert Healy. 1990. *Natural Resource Policymaking in Developing Countries*. Durham, N.C.: Duke University Press.

Asian Development Bank. 1973. Proceedings of Regional Workshop on Irrigation Water Management. Manila: Asian Development Bank.

Aumann, R. J. 1976. "Agreeing to Disagree." *The Annals of Statistics* 4(1): 236–39.

Axelrod, R. 1981. "The Emergence of Cooperation among Egoists." *American Political Science Review* 75: 306–18.

———. 1984. *The Evolution of Cooperation*. New York: Basic Books.

Bagadion, Benjamin U. 1988. "The Evolution of the Policy Context: An Historical Overview." In *Transforming a Bureaucracy: The Experience of the Philippine National Irrigation Administration,* ed. Frances F. Korten and Robert Y. Siy, Jr., 1–19. West Hartford, Conn.: Kumarian Press.

Barker, Randolph, E. Walter Coward, Jr., Gilbert Levine, and Leslie E. Small. 1984. *Irrigation Development in Asia: Past Trends and Future Directions*. Ithaca, N.Y.: Cornell University Press.

Barnett, Tony. 1977. *The Gezira Scheme: An Illusion of Development*. London: Frank Cass.

Bates, Robert. 1987. *Essays on the Political Economy of Rural Africa*. Berkeley: University of California Press.

Benedict, Peter, Ahmed H. Ahmed, Rollo Ehrich, Stephen F. Lintner, Jack Morgan, and Mohamed A. M. Salih. 1982. *Sudan: The Rahad Irrigation Project*. Project Impact Evaluation no. 31. Washington, D.C.: U.S. Agency for International Development.

Binswanger, Hans, and Prabhu Pingali. 1988. "Technological Priorities for Farming in Sub-Saharan Africa." *World Bank Research Observer* 3 (1).

Blomquist, William. (forthcoming). *They Prefer Chaos*. San Francisco: Institute for Contemporary Studies Press.

Bottrall, Anthony. 1981. *Comparative Study of the Management and Organization of Irrigation Projects*. Staff Working Paper no. 458. Washington, D.C.: World Bank.

Boudreaux, D. J., and R. G. Holcombe. 1989. "Government by Contract." *Public Finance Quarterly* 17: 264–80.

Breton, Albert, and Ronald Wintrobe. 1981. *The Logic of Bureaucratic Conduct*. Cambridge: Cambridge University Press.

Bromley, Daniel W. 1982. "Improving Irrigated Agriculture: Institutional Reform and the Small Farmer." Staff Working Paper no. 531. Washington, D.C.: World Bank.

Brown, L. David, and David C. Korten. 1989. *Understanding Voluntary Organizations*. PPR Working Paper Series 258. Washington, D.C.: World Bank.

Buchanan, James M., Robert D. Tollison, and Gordon Tullock, eds. 1980. *Toward a Theory of the Rent-Seeking Society*. College Station: Texas A&M University Press.

Byrne, J. A. 1986. "The Decline in Paddy Cultivation in a Dry Zone Village of Sri Lanka." In *Rice Societies: Asian Problems and Prospects*, ed. Irene Norlund, Sven Cederroth, and Ingela Gerdin, 81–116. London: Curzon Press.

Carruthers, Ian. 1981. "Neglect of O&M in Irrigation, the Need for New Sources and Forms of Support." *Water Supply and Management* 5: 53–65.

———. 1988. "Irrigation under Threat: A Warning Brief for Irrigation Enthusiasts." *IIMI Review* 2: 8–11, 24–25.

Cernea, Michael M., ed. 1985. *Putting People First: Sociological Variables in Rural Development*. New York: Oxford University Press.

———. 1987. "Farmer Organization and Institution Building for Sustainable Development." *Regional Development Dialogue* 8(2): 1–24. Nagoya, Japan: United Nations Centre for Regional Development.

Chambers, Robert. 1980. "Basic Concepts in the Organization of Irrigation." In *Irrigation and Agricultural Development in Asia: Perspectives from the Social Sciences,* ed. E. Walter Coward, Jr., 28–50. Ithaca, N.Y.: Cornell University Press.

———. 1988. *Managing Canal Irrigation: Practical Analysis from South Asia.* Cambridge: Cambridge University Press.

Ciriacy-Wantrup, S. V., and R. C. Bishop. 1975. " 'Common Property' as a Concept in Natural Resource Policy." *Natural Resources Journal* 15: 713–27.

Coleman, James. 1986. "Social Theory, Social Research, and a Theory of Action." *American Journal of Sociology* 91(1): 309–35.

———.1988. "Social Capital in the Creation of Human Capital." *American Journal of Sociology* 94 (supplement): S95–S120.

Colmey, John. 1988. "Irrigated Non-Rice Crops: Asia's Untapped Resource." *IIMI Review* 2(1): 3–7.

Commons, John R. 1957. *Legal Foundations of Capitalism.* Madison: University of Wisconsin Press.

Corey, A. T. 1986. Control of Water within Farm Turnouts in Sri Lanka. In *Proceedings of a Workshop on Water Management in Sri Lanka.* Documentation Series no. 10. Colombo: Agrarian Research and Training Institute.

Coward, E. Walter, Jr. 1979. "Principles of Social Organization in an Indigenous Irrigation System." *Human Organization* 38(1): 28–36.

———, ed. 1980. *Irrigation and Agricultural Development in Asia: Perspectives from the Social Sciences.* Ithaca, N.Y.: Cornell University Press.

———. 1985. "Technical and Social Change in Currently Irrigated Regions: Rules, Roles, and Rehabilitation." In *Putting People First: Sociological Variables in Rural Development,* ed. Michael M. Cernea, 27–52. Oxford: Oxford University Press.

Craven, K., J. Merryman, and N. Merryman. 1989. *Jubba Environmental and Socioeconomic Studies.* Burlington, Vt.: Associates in Rural Development.

Crosson, Pierre R., and Norman J. Rosenberg. 1989. "Strategies for Agriculture." *Scientific American* 261: 128–35.

David, Paul A. 1988. "Path Dependence: Putting the Past into the Future of Economics." Working paper. Stanford: Stanford University Department of Economics.

de los Reyes, Romana P., and Sylvia Ma. G. Jopillo. 1988. "The Impact of Participation: An Evaluation of the NIA's Communal Irrigation

Program." In *Transforming a Bureaucracy: The Experience of the Philippine National Irrigation Administration*, ed. Frances F. Korten and Robert Y. Siy, Jr., 90–116. West Hartford, Conn.: Kumarian Press.

de Soto, Hernando. 1989. *The Other Path. The Invisible Revolution in the Third World*. New York: Harper & Row.

Dhawan, B. D. 1988. *Irrigation in India's Agricultural Development: Productivity, Stability, Equity*. New Delhi: Sage.

Dosi, Giovanni. 1988. "Technical Change, Institutional Processes and Economic Dynamics: Some Tentative Propositions and a Research Agenda." Rome: University of Rome Department of Economics.

Downing, Theodore E. 1974. "Irrigation and Moisture-Sensitive Periods: A Zapotec Case." In *Irrigator's Impact on Society*, ed. Theodore E. Downing and McGuire Gibson, 113–22. Tuscon: University of Arizona Press.

Easter, K. William. 1985. *Recurring Costs of Irrigation in Asia: Operation and Maintenance*. Ithaca, N.Y.: Cornell University Water Management Synthesis Project.

Freeman, David M., and Max L. Lowdermilk. 1985. "Middle-level Organizational Linkages in Irrigation Projects." In *Putting People First: Sociological Variables in Rural Development,* ed. Michael M. Cernea, 91–118. Oxford: Oxford University Press.

Geertz, Clifford. 1980. "Organization of the Balinese Subak." In *Irrigation and Agricultural Development in Asia: Perspectives from the Social Sciences,* ed. E. Walter Coward, Jr., 70–90. Ithaca, N.Y.: Cornell University Press.

General Accounting Office. 1983. *Irrigation Assistance to Developing Countries Should Require Stronger Commitments to Operation and Maintenance*. Gaithersburg, Md.: General Accounting Office.

Gillespie, Victor A. 1975. *Farmer Irrigation Associations and Farmer Cooperation*. East-West Food Institute Paper no. 3. Honolulu: Food Institute, East-West Center.

Groenfeldt, David, and Joyce L. Moock, eds. 1989. *Social Science Perspectives on Managing Agricultural Technology*. Colombo, Sri Lanka: International Irrigation Management Institute.

Harriss, John C. 1984. "Social Organization and Irrigation: Ideology, Planning and Practice in Sri Lanka's Settlement Schemes." In *Understanding Green Revolutions,* ed. T. P. Bayliss-Smith and S. Wanmali, 315–38. Cambridge: Cambridge University Press.

Hayek, F. A. 1945. "The Use of Knowledge in Society." *American Economic Review* 35(4): 519–30.

Hechter, Michael. 1987. *Principles of Group Solidarity*. Berkeley: University of California Press.

Heckathorn, D. D. 1984. "A Formal Theory of Social Exchange: Process and Outcome." *Current Perspectives in Social Theory* 5: 145–80.

Hilton, Rita M. 1990. Cost Recovery and Local Resource Mobilization in Irrigation Systems in Nepal: Case Study of Karjahi Irrigation System. Paper presented at the annual meeting of the International Association for the Study of Common Property, Duke University, September 28–30.

Hunt, Robert C. 1989. "Appropriate Social Organization? Water User Associations in Bureaucratic Canal Irrigation Systems." *Human Organization* 48(1): 79–90.

———. 1990. "Organizational Control over Water: The Positive Identification of a Social Constraint on Farmer Participation." In *Social, Economic, and Institutional Issues in Third World Irrigation Management* ed. R. K. Sampath and Robert A. Young, 141–54. Studies in Water Policy and Management no. 15. Boulder, Colo.: Westview Press.

International Bank for Reconstruction and Development. 1985. *Tenth Annual Review of Project Management in Sri Lanka's Irrigation Schemes 1984*. Washington, D.C.: World Bank Operations Evaluation Department.

Jayawardene, Jayantha. 1986. "The Training of Mahaweli Turnout Group Leaders." In *Participatory Management in Sri Lanka's Irrigation Schemes*, 77–85. Digana Village, Sri Lanka: International Irrigation Management Institute.

Kaye, Lincoln. 1989. "The Wasted Waters." *Far Eastern Economic Review* 143(5) (2 February), 16–22.

Kiser, Larry L., and Elinor Ostrom. 1982. "The Three Worlds of Action: A Metatheoretical Synthesis of Institutional Approaches." In *Strategies of Political Inquiry,* ed. E. Ostrom, 179–222. Beverly Hills: Sage.

Korten, David C. 1980. "Community Organization and Rural Development: A Learning Process Approach." *Public Administration Review* 40(5): 480–511.

———. 1988. "From Bureaucratic to Strategic Organization." In *Transforming a Bureaucracy: The Experience of the Philippine National Irrigation Administration,* ed. Frances F. Korten and Robert Y. Siy, Jr., 117–44. West Hartford, Conn.: Kumarian Press.

Korten, Frances F., and Robert Y. Siy, Jr., eds. 1988. *Transforming a Bureaucracy: The Experience of the Philippine National Irrigation Administration*. West Hartford, Conn.: Kumarian Press.

Krueger, Anne O. 1974. "The Political Economy of the Rent-Seeking Society." *American Economic Review* 64: 291–301.

Lachmann, Ludwig M. 1978. *Capital and Its Structure*. Kansas City: Sheed Andrews & McMeel.

Leach, E. R. 1959. "Hydraulic Society in Ceylon." *Past and Present* 15: 2–26.

Levi, Margaret. 1988. *Of Rule and Revenue*. Berkeley: University of California Press.

Levine, Gilbert. 1980. "The Relationship of Design, Operation and Management." In *Irrigation and Agricultural Development in Asia: Perspectives from the Social Sciences,* ed. E. Walter Coward, Jr., 51–62. Ithaca, N.Y.: Cornell University Press.

Lewis, T. R., and J. Cowens. 1983. "Cooperation in the Commons: An Application of Repetitious Rivalry." Vancouver: University of British Columbia Department of Economics.

Maass, A., and R. L. Anderson. 1986. *. . . and the Desert Shall Rejoice: Conflict, Growth and Justice in Arid Environments.* Malabar, Fla.: Robert E. Krieger.

Madduma Bandara, C. M. 1977. "Hydrological Consequences of Agrarian Change." In *Green Revolution? Technology and Change in Rice Growing Areas of Tamil Nadu and Sri Lanka,* ed. B. H. Farmer. New York: Macmillan.

Martin, Edward G. 1986. "Resource Mobilization, Water Allocation, and Farmer Organization in Hill Irrigation Systems in Nepal." Ph.D. diss., Cornell University.

Martin, Edward G., and Robert Yoder. 1983. The Chherlung Thulo Kulo: A Case Study of a Farmer-Managed Irrigation System. In *Water Management in Nepal: Proceedings of the Seminar on Water Management Issues, July 31–August 2*, Appendix I, 203–17. Kathmandu, Nepal: Ministry of Agriculture, Agricultural Projects Services Centre, and the Agricultural Development Council.

Mehra, S. 1981. *Instability in Indian Agriculture in the Context of the New Technology.* Research Report no. 25. Washington, D.C.: International Food Policy Research Institution.

Moore, Mick. 1989. "The Fruits and Fallacies of Neoliberalism: The Case of Irrigation Policy." *World Politics* 17(1): 733–50.

Moris, Jon, and Derrick J. Thom. 1990. *Irrigation Development in Africa. Lessons of Experience.* Studies in Water Policy and Management no. 14. Boulder, Colo.: Westview Press.

Nelson, Richard R., and Sidney G. Winter. 1982. *An Evolutionary Theory of Economic Change.* Cambridge, Mass.: Harvard University Press.

Niasse, Madiodio. 1990. "Village Irrigated Perimeters at Doumga Rindiaw, Senegal." *Development Anthropology Network* 8 (Spring): 6–11. Binghamton, N.Y.: Institute for Development Anthropology.

North, Douglass C. 1989. "Institutions and Economic Growth: An Historical Introduction." *World Development* 17(9): 319–32.

Nyoni, Sithembiso. 1987. "Indigenous NGOs: Liberation, Self-reliance, and Development." *World Development,* 15(Supplement): 51–56.

Oakerson, Ronald J. 1986. A Model for the Analysis of Common Property Problems. In *Proceedings of the Conference on Common Property Resource Management,* National Research Council, 13–30. Washington, D.C.: National Academy Press.

———. 1988 "Reciprocity: A Bottom-Up View of Political Development." In *Rethinking Institutional Analysis and Development: Issues, Alternatives, and Choices,* ed. Vincent Ostrom, David Feeny, and Hartmut Picht, 141–58. San Francisco: Institute for Contemporary Studies Press.

Olson, Mancur. 1965. *The Logic of Collective Action: Public Goods and the Theory of Groups*. Cambridge: Harvard University Press.

———.1969. "The Principle of 'Fiscal Equivalence': The Division of Responsibilities among Different Levels of Government." *American Economic Review* 59(2): 479–87.

O'Mara, Gerald T. 1989. "Issues and Options in Irrigation Finance." In *Innovation in Resource Management: Proceedings of the Ninth Agriculture Sector Symposium,* ed. L. Richard Meyers. Washington, D.C.: World Bank.

Ostrom, Elinor. 1986. "An Agenda for the Study of Institutions." *Public Choice* 48: 3–25.

———.1990. *Governing the Commons: The Evolution of Institutions for Collective Action*. New York: Cambridge University Press.

Ostrom, Elinor, Larry Schroeder, and Susan Wynne. 1990. *Institutional Incentives and Rural Infrastructure Sustainability.* Burlington, Vt.: Associates in Rural Development.

Ostrom, Vincent. 1976. "The Contemporary Debate over Centralization and Decentralization." *Publius* 6(4): 21–32.

———. 1980. "Artisanship and Artifact." *Public Administration Review* 40: 309–17.

———. 1982. "A Forgotten Tradition: The Constitutional Level of Analysis." In *Missing Elements in Political Inquiry: Logic and Levels of Analysis,* ed. Judith A. Gillespie and Dina A. Zinnes, 237–52. Beverly Hills: Sage.

———.1986. "A Fallabilist's Approach to Norms and Criteria of Choice." In *Guidance, Control, and Evaluation in the Public Sector,* ed. Franz-Xaver Kaufmann, Giandomenico Majone, and Vincent Ostrom, 229–49. Berlin and New York: Walter de Gruyter.

———. 1987. *The Political Theory of a Compound Republic: Designing the American Experiment*. 2d ed. Lincoln: University of Nebraska Press.

———. 1988. "Cryptoimperialism, Predatory States, and Self-Governance." In *Rethinking Institutional Analysis and Development: Issues, Alternatives, and Choices,* ed. Vincent Ostrom, David Feeny, and Hartmut Picht, 43–68. San Francisco: Institute for Contemporary Studies Press.

———. 1989. *The Intellectual Crisis in American Public Administration*. 2d rev. ed. Lincoln: University of Nebraska Press.

———. 1991. The *Meaning of American Federalism: Constituting a Self-Governing Society*. San Francisco: Institute for Contemporary Studies Press.

Ostrom, Vincent, David Feeny, and Hartmut Picht, eds. 1988. *Rethinking Institutional Analysis and Development: Issues, Alternatives, and Choices*. San Francisco: Institute for Contemporary Studies Press.

Ostrom, Vincent, and Elinor Ostrom. 1978. "Public Goods and Public Choices." In *Alternatives for Delivering Public Services: Toward Improved Performance,* ed. E. S. Savas, 7–49. Boulder, Colo.: Westview Press.

Pant, Niranjan, ed. 1984. *Productivity and Equity in Irrigation Systems*. New Delhi: Ashish.

Patil, R. K. 1986. Pani Panchayats in Mula Command, Ahmednagar District, Maharashtra State. Paper for the Symposium on Community Irrigation Systems organized by the National Institute of Bank Management.

Plusquellec, Herve L., 1989. *Two Irrigation Systems in Colombia: Their Performance and Transfer of Management to Users' Associations*. PPR Working Paper Series 264. Washington, D.C.: World Bank.

———. 1990. *The Gezira Irrigation Scheme in Sudan. Objectives, Design, and Performance*. Technical Paper no. 120. Washington, D.C.: World Bank.

Plusquellec, Herve L., and Thomas H. Wickham. 1985. *Irrigation Design and Management: Experience in Thailand and Its General Applicability.* Technical Paper no. 40. Washington, D.C.: World Bank.

Pradhan, Prachanda. 1989a. *Patterns of Irrigation Organization in Nepal*. Colombo, Sri Lanka: International Irrigation Management Institute.

———. 1989b. *Increasing Agricultural Production in Nepal: Role of Low-Cost Irrigation Development through Farmer Participation*. Kathmandu, Nepal: International Irrigation Management Institute.

Reidinger, Richard B. 1974. "Institutional Rationing of Canal Water in Northern India: Conflict between Traditional Patterns and Modern Needs." *Economic Development and Cultural Change* 23(1): 79–104.

Repetto, Robert. 1986. *Skimming the Water: Rent-Seeking and the Performance of Public Irrigation Systems*. Research Report no. 41. Washington, D.C.: World Resources Institute.

Riordon, William L. 1963. *Plunkitt of Tammany Hall*. New York: E. P. Dutton.

Roy, S. K. 1979. "Irrigation Development under India's New Plan (1978–83): An Appraisal." In *Agricultural Situation in India,* India, Ministry of Agriculture and Irrigation. New Delhi: Manager of Publications, Civil Lines.

Runge, C. F. 1986. "Common Property and Collective Action in Economic Development." In *Proceedings of the Conference on Common Property Resource Management,* ed. National Research Council, 31–60. Washington, D.C.: National Academy Press.

Sampath, R. D., and Robert A. Young, eds. 1990. *Social, Economic, and Institutional Issues in Third World Irrigation Management*. Studies in Water Policy and Management no. 15. Boulder, Colo.: Westview Press.

Scharpf, Fritz W. 1990. "Games Actors May Play: The Problem of Predictability." *Rationality and Society* 2(4): 471–94.

Scudder, Thayer. 1990. "Victims of Development Revisited: The Political Costs of River Basin Development." *Development Anthropology Network* 8 (Spring): 1–5. Binghamton, N.Y.: Institute for Development Anthropology.

Shah, Tushaar. 1985. *Transforming Ground Water Markets into Powerful Instruments of Small Farmer Development: Lessons from the Punjab, Uttar Pradesh, and Gujarat*. Anand, India: Institute of Rural Management.

———. 1986. *Optimal Ground Water Markets: A Theoretical Framework*. Anand, India: Institute of Rural Management.

Sharan, Girja, and S. Narayanan. 1983. *Duration of Stay and Frequency of Transfers of District and Lower Level Officials*. The Rural University Experiment II: Occasional Paper 1. Ahmedabad: Indian Institute of Management.

Shivakoti, Ganesh, and Khadka Giri. 1990. "Effects of Different Types and Levels of Intervention in Farmer Managed Irrigation Systems in Nepal." Draft Report for Irrigation Management Project, Government of Nepal.

Simon, Herbert A., Donald W. Smithburg, and Victor A. Thompson. 1958. *Public Administration*. New York: Alfred A. Knopf.

Singh, K. K. 1983. "Farmers' Organisation and Warabandi in the Sriramasagar (Pochampad) Project." In *Utilization of Canal Waters: A Multi-disciplinary Perspective on Irrigation,* Publication No. 164, ed. K. K. Singh, 97–101. New Delhi: Central Board for Irrigation and Power.

Siy, Robert Y., Jr. 1982. *Community Resource Management: Lessons from the Zanjera*. Quezon City: University of the Philippines Press.

———. 1988. "A Tradition of Collective Action: Farmers and Irrigation in the Philippines." In *Transforming a Bureaucracy: The Experience of the Philippine National Irrigation Administration,* ed. Frances F. Korten and Robert Y. Siy, Jr. West Hartford, Conn.: Kumarian Press.

Small, Leslie, Marietta Adriano, and Edward D. Martin. 1986. *Regional Study on Irrigation Service Fees: Final Report*. Kandy, Sri Lanka: International Irrigation Management Institute.

Sommer, J. G., R. Aquino, C. A. Fernandez, F. H. Golay, and E. Simmons. 1982. *Philippines: BICOL Integrated Area Development*. Project Impact Evaluation Report no. 28. Washington, D.C.: U.S. Agency for International Development.

Steinberg, David I., et al. 1983. *Irrigation and AID's Experience: A Consideration Based on Evaluations.* AID Program Evaluation Report no. 8. Washington, D.C.: U.S. Agency for International Development.

Sugden, R. 1986. *The Economics of Rights, Co-operation and Welfare.* Oxford: Basil Blackwell.

Tang, Shui Yan. 1991. "Institutional Arrangements and the Management of Common-Pool Resources." *Public Administration Review* 51(1): 42–51.

———. 1992. *Institutions and Collective Action: Self-Governance in Irrigation.* San Francisco: Institute for Contemporary Studies Press.

Tendler, Judith. 1975. *Inside Foreign Aid.* Baltimore, Md.: Johns Hopkins University Press.

Tollison, Robert B. 1982. "Rent Seeking: A Survey." *Kyklos* 35(4): 575–602.

Tsebelis, George. 1989. "The Abuse of Probability in Political Analysis: The Robinson Crusoe Fallacy." *American Political Science Review* 83: 77–91.

———. 1990. *Nested Games: Political Context, Political Institutions and Rationality.* Berkeley: University of California Press.

Uphoff, Norman. 1985. "People's Participation in Water Management: Gal Oya, Sri Lanka." In *Public Participation in Development Planning and Management: Cases from Africa and Asia,* ed. J. C. Garcia-Zamor, 131–78. Boulder, Colo.: Westview Press.

———. 1986. *Improving International Irrigation Management with Farmer Participation: Getting the Process Right.* Boulder, Colo.: Westview Press.

Uphoff, Norman, M. L. Wickramasinghe, and C. M. Wijayaratna. 1990. " 'Optimum' Participation in Irrigation Management: Issues and Evidence from Sri Lanka." *Human Organization* 49(1): 26–40.

U.S. Agency for International Development. 1983. *Irrigation and AID's Experience: A Consideration Based on Evaluations.* AID Program Evaluation Report no. 8. Washington, D.C.: U.S. Agency for International Development.

Vander Velde, Edward J. 1980. "Local Consequences of a Large-Scale Irrigation System in India." In *Irrigation and Agricultural Development in Asia,* ed. E. Walter Coward, Jr., 199–328. Ithaca, N.Y.: Cornell University Press.

Wade, Robert. 1982a. *Irrigation and Agricultural Politics in South Korea.* Boulder, Colo.: Westview Press.

———. 1982b. "The System of Administrative and Political Corruption: Canal Irrigation in South India." *Journal of Development Studies* 18(3): 287–328.

———. 1982c. "Corruption: Where Does the Money Go?" *Economic and Political Weekly* 17(40): 1606.

———. 1985. "The Market for Public Office: Why the Indian State Is Not Better at Development." *World Development* 13(4): 467–97.

———. 1987. "Managing Water Managers: Deterring Expropriation or Equity as a Control Mechanism." In *Water and Water Policy in World Food Supplies,* ed. Wayne R. Jordon, 117–83. College Station: Texas A&M University Press.

———. 1988. *Village Republics: Economic Conditions for Collective Action in South India.* Cambridge: Cambridge University Press.

———. n.d. "The Evolution of Water Rights in South India: Directed and Induced." Washington, D.C.: World Bank. Mimeo.

Weissing, Franz, and Elinor Ostrom. 1991. "Irrigation Institutions and the Games Irrigators Play: Rule Enforcement Without Guards." In *Game Equilibrium Models II: Methods, Morals, and Markets*, ed. Reinhard Selten, 188–262. Berlin: Springer-Verlag.

Williamson, Oliver E. 1979. "Transaction Cost Economics: The Governance of Contractual Relations." *Journal of Law and Economics* 22(2): 233–61.

———. 1985. *The Economic Institutions of Capitalism: Firms, Markets, Relational Contracting.* New York: Free Press.

Wittfogel, Karl A. 1957. *Oriental Despotism.* New Haven: Yale University Press.

Wolf, Edward C. 1986. "Beyond the Green Revolution: New Approaches for Third World Agriculture." *World Watch Paper* 73 (October).

Wunsch, James, and Dele Olowu, eds. 1990. *The Failure of the Centralized State: Institutions and Self-Governance in Africa.* Boulder, Colo.: Westview Press.

Yudelman, Montague. 1985. *The World Bank and Agricultural Development: An Insider's View.* World Resources Paper no. 1. Washington, D.C.: World Resources Institute.

———. 1987. "The World Bank and Irrigation." In *Water and Water Policy in World Food Supplies: Proceedings of the Conference, 26–30 May 1985,* 419–23. College Station: Texas A&M University.

———. 1989. "Sustainable and Equitable Development in Irrigation Environments." In *Environment and the Poor: Development Strategies for a Common Agenda,* ed. W. Jeffrey Leonard, 61–85. New Brunswick: Transaction Books.

ORDER FORM

Please accept this order for the following book:

	Qty	*Price*	*Total*
Ostrom, **Crafting Institutions for Self-Governing Irrigation Systems** (paper)	_____	x $9.95 =	_____
Tang, **Instititutions and Collective Action: Self-Governance in Irrigation** (paper)	_____	x $9.95 =	_____
		Subtotal	_____
		Shipping charges	_____
CA residents please add applicable state and local sales tax			_____
		TOTAL	_____

Shipping Charges: In North America, $3.00 for first book, $.75 for each additional book. Outside North America, $3.00 per book surface mail, $10.00 per book air mail.

Please include payment with order or provide full credit card information. Personal checks are accepted when drawn in U.S. funds on a U.S. bank.

Name ____

Institution ____________________________________

Address ______________________________________

City ______________ State ______ Zip ________

Country ________________________

❒ Check enclosed ❒ MasterCard ❒ VISA

Credit Card # ___________________________________

Expires __________ Signature ____________________

❒ Please add my name to the ICS Press mailing list.

MAIL orders to: **ICS Press, 243 Kearny St., San Francisco, CA 94108**

FAX orders to : **(415) 986-4878**

PHONE orders to: **(800) 326-0263 toll free** in the U.S., or **(415) 981-5353**

Quantity Discounts are available; please call (800) 326-0263 for details.